The IMA Volumes
in Mathematics
and its Applications

Volume 147

Series Editors
Douglas N. Arnold   Arnd Scheel

Institute for Mathematics and
its Applications (IMA)

The Institute for Mathematics and its Applications was established by a grant from the National Science Foundation to the University of Minnesota in 1982. The primary mission of the IMA is to foster research of a truly interdisciplinary nature, establishing links between mathematics of the highest caliber and important scientific and technological problems from other disciplines and industries. To this end, the IMA organizes a wide variety of programs, ranging from short intense workshops in areas of exceptional interest and opportunity to extensive thematic programs lasting a year. IMA Volumes are used to communicate results of these programs that we believe are of particular value to the broader scientific community.

The full list of IMA books can be found at the Web site of the Institute for Mathematics and its Applications:

http://www.ima.umn.edu/springer/volumes.html

Presentation materials from the IMA talks are available at

http://www.ima.umn.edu/talks/

Douglas N. Arnold, Director of the IMA

* * * * * * * * * *

IMA ANNUAL PROGRAMS

1982–1983	Statistical and Continuum Approaches to Phase Transition
1983–1984	Mathematical Models for the Economics of Decentralized Resource Allocation
1984–1985	Continuum Physics and Partial Differential Equations
1985–1986	Stochastic Differential Equations and Their Applications
1986–1987	Scientific Computation
1987–1988	Applied Combinatorics
1988–1989	Nonlinear Waves
1989–1990	Dynamical Systems and Their Applications
1990–1991	Phase Transitions and Free Boundaries
1991–1992	Applied Linear Algebra
1992–1993	Control Theory and its Applications
1993–1994	Emerging Applications of Probability
1994–1995	Waves and Scattering
1995–1996	Mathematical Methods in Material Science
1996–1997	Mathematics of High Performance Computing

(Continued at the back)

Grzegorz A. Rempała   Jacek Wesołowski
Authors

Symmetric Functionals on Random Matrices and Random Matchings Problems

 Springer

Grzegorz A. Rempała
Department of Mathematics
University of Louisville, KY, USA
Louisville 40292
http://www.louisville.edu/~garemp01

Jacek Wesołowski
Wydzial Matematyki i Nauk
Informacyjnych
Politechnika Warszawska, Warszawa,
1 Pl. Politechniki
Warszawa 00-661
Poland

ISBN: 978-0-387-75145-0   e-ISBN: 970-0-387-75146-7

Mathematics Subject Classification (2000): 60F05, 60F17, 62G20, 62G10, 05A16

Library of Congress Control Number: 2007938212

Printed on acid-free paper.

9 8 7 6 5 4 3 2 1

springer.com

Moim Rodzicom, Helenie oraz Jasiowi i Antosiowi
(Grzegorz Rempała)

Moim Rodzicom
(Jacek Wesołowski)

Foreword

This IMA Volume in Mathematics and its Applications

**SYMMETRIC FUNCTIONALS ON RANDOM MATRICES
AND RANDOM MATCHINGS PROBLEMS**

During the academic year 2003–2004, the Institute for Mathematics and its Applications (IMA) held a thematic program on Probability and Statistics in Complex Systems. The program focused on large complex systems in which stochasticity plays a significant role. Such systems are very diverse, and the IMA program emphasized systems as varied as the human genome, the internet, and world financial markets. Although quite different, these systems have features in common, such as multitudes of interconnecting parts and the availability of large amounts of high-dimensional noisy data. The program emphasized the development and application of common mathematical and computational techniques to model and analyze such systems. About 1,000 mathematicians, statisticians, scientists, and engineers participated in the IMA thematic program, including about 50 who were in residence at the IMA during much or all of the year.

The present volume was born during the 2003–2004 thematic program at the IMA. The two authors were visitors to the IMA during the program year, with the first author resident for the entire ten months. This volume is a result of the authors' interactions at the IMA, and, especially their discussions with the many other program participants in the program and their involvement in the numerous tutorials, workshops, and seminars held during the year. The book treats recent progress in random matrix permanents, random matchings and their asymptotic behavior, an area of stochastic modeling and analysis which has applications to a variety of complex systems and problems of high dimensional data analysis.

Like many outcomes of IMA thematic programs, the seed for this volume was planted at the IMA, but it took time to grow and flourish. The final fruit is realized well after the program ends. While all credit and responsibility for the contents of the book reside with the authors, the IMA is delighted to have supplied the fertile ground for this work to take place.

We take this opportunity to thank the National Science Foundation for its support of the IMA.

Series Editors

Douglas N. Arnold, Director of the IMA

Arnd Scheel, Deputy Director of the IMA

Preface

The idea of writing this monograph came about through discussions which we held as participants in the activities of an annual program "Probability and Statistics in Complex Systems" of the Institute for Mathematics and Its Applications at the University of Minnesota (IMA) which was hosted there during the 2003/04 academic year. In the course of interactions with the Institute's visitors and guests, we came to a realization that many of the ideas and techniques developed recently for analyzing asymptotic behavior of random matchings are relatively unknown and could be of interest to a broader community of researchers interested in the theory of random matrices and statistical methods for high dimensional inference. In our IMA discussions it also transpired that many of the tools developed for the analysis of asymptotic behavior of random permanents and the likes may be also useful in more general context of problems emerging in the area of complex stochastic systems. In such systems, often in the context of modeling, statistical hypothesis testing or estimation of the relevant quantities, the distributional properties of the functionals on the entries of random matrices are of concern. From this viewpoint, the interest in the laws of various random matrix functionals useful in statistical analysis contrasts with the interest of a classical theory of random matrices which is primarily concerned with asymptotic distributional laws of eigenvalues and eigenvectors.

The text's content is drawn from the recent literature on questions related to asymptotics for random permanents and random matchings. That material has been augmented with a sizable amount of preliminary material in order to make the text somewhat self-contained. With this supplementary material, the text should be accessible to any mathematics, statistics or engineering graduate student who has taken basic introductory courses in probability theory and mathematical statistics.

The presentation is organized in seven chapters. Chapter 1 gives a general introduction to the topics covered in the text while also providing the reader with some examples of their applications to problems in stochastic complex systems formulated in terms of random matchings. This preliminary chapter makes a connection between random matchings, random permanents and U-statistics. Also a concept of a P- statistic, which connects the three concepts is introduced there. Chapter 2 builds upon these connections and contains a number of results for a general class of random matchings which, like for instance the variance formula for a P-statistic, are fundamental to the developments further in the text. Taken together the material of Chapters 1 and 2 should give the reader the necessary background to approach the topics covered later in the text.

Chapters 3 and 4 deal with random permanents and a problem of describing asymptotic distributions for a "classical" count of perfect matchings in random bipartite graphs. Chapter 3 details a relatively straightforward but computationally tedious approach leading to central limit theorems and laws of large numbers for random permanents. Chapter 4 presents a more general treatment of the subject by means of functional limit theorems and weak convergence of iterative stochastic integrals. The basic facts of the theory of stochastic integration are outlined in the first sections of Chapter 4 as necessary.

In Chapter 5 the results on asymptotics of random permanents are extended to P-statistics, at the same time covering a large class of matchings. The limiting laws are expressed with the help of multiple Wiener-Itô integrals. The basic properties of a multiple Wiener-Itô integral are summarized in the first part of the chapter. Several applications of the asymptotic results to particular counting problems introduced in earlier chapters are presented in detail.

Chapter 6 makes a connection between P-statistics and matchings on one side and the "incomplete" U-statistics on the other. The incomplete permanent design is analyzed first. An overview of the analysis of both asymptotic and finite sample properties of P-statistics in terms of their variance efficiency as compared with the corresponding "complete" statistics is presented. In the second part minimum rectangular designs (much lighter that permanent designs) are introduced and their efficiency is analyzed. Also their relations to the concept of mutual orthogonal Latin squares of classical statistical design theory is discussed there.

Chapter 7 covers some of the recent results on the asymptotic lognormality of sequences of products of increasing sums of independent identically distributed random variables and their U-statistics counterparts. The developments of the chapter lead eventually to a limit theorem for random determinants for Wishart matrices. Here again, similarly as in some of the earlier-discussed limit theorems for random permanents, the lognormal law appears in the limit.

We would like to express our thanks to several individuals and institutions who helped us in completing this project. We would like to acknowledge the IMA director, Doug Arnold who constantly encouraged our efforts, as well as our many other colleagues, especially André Kézdy, Ofer Zeitouni and Shmuel Friedland, who looked at and commented on the various finished and not-so-finished portions of the text. We would also like to thank Ewa Kubicka and Grzegorz Kubicki for their help with drawing some of the graphs presented in the book. Whereas the idea of writing the current monograph was born at the IMA, the opportunity to do so was also partially provided by other institutions. In particular, the Statistical and Applied Mathematical Sciences Institute in Durham, NC held during 2005/6 academic year a program on "Random Matrices and High Dimensional Inference" and kindly invited the first of the authors to participate in its activities as a long term visitor. The project was also supported by local grants from the Faculty of Mathematics and Information Science of the Warsaw University of Technology, Warszawa, Poland and from the Department of Mathematics at the University of Louisville.

Louisville, KY and Warszawa (Poland)   *Grzegorz A. Rempała*
July 2007   *Jacek Wesołowski*

Contents

1

Basic Concepts

1.1 Bipartite Graphs in Complex Stochastic Systems

Ever since its introduction in the seminal article of Erdős and Rényi (1959) a random graph has become a very popular tool in modeling complicated stochastic structures. In particular, a *bipartite random graph* has become an important special case arising naturally in many different application areas amenable to the analysis by means of random-graph models. The use of these models have been steadily growing over the last decade as the rapid advances in scientific computing made them more practical in helping analyze a wide variety of complex stochastic phenomena, in such diverse areas as the internet traffic analysis, the biochemical (cellular) reactions systems, the models of social interactions in human relationships, and many others. In modern terms these complex stochastic phenomena are often referred to as the stochastic *complex systems* and are typically characterized by the very large number of nonlinearly interacting parts that prevent one from simply extrapolating overall system behavior from its isolated components.

For instance, in studying internet traffic models concerned with subnetworks of computers communicating only via the designated "gateway" sets, such sets of gateways are often assumed to exchange data packets directly only between (not within) their corresponding sub-networks. These assumptions give rise to graphical representations via bipartite (random) graphs, both directed and undirected, which are used to develop computational methods to determine the properties of such internet communication networks.

Similarly, in an analysis of a system of biochemical interactions aiming at modeling of the complicated dynamics of living cells, a bipartite graph arises naturally as a model of biochemical reactions where the two sets of nodes represent, respectively, the chemical molecular species involved and the chemical interaction types among them (i.e., chemical reaction channels). A directed edge from a species node to a reaction node indicated that species is a reactant for that node, and an edge from a reactions node to a species one, indicates that the species is a product of the reaction. The so called reaction stoichiometric constants may be used as weights associated with each edge, so the chemical reaction graph is typically a directed, weighted graph. The graph is also random if we assume, as is often done in Bayesian models analysis, some level of uncertainty about the values of the stoichiometric constants. Such graphs are closely related to the Petri net models studied in computer science in connection with a range of modeling problems (see, e.g., Peterson, 1981).

In the area of sociology known as social networks analysis, the so called affiliation network is a simplest example of a bipartite graph where one set of nodes represents all of the individuals in the network and another one all available affiliations. The set of undirected edges between the individual nodes and the affiliation nodes represents then membership structure where two individuals are associated with each other if they have a common affiliation node. Examples that have been studied in this context include for instance networks of individuals joined together by common participation in social events and CEOs of companies joined by common membership of social clubs. The collaboration networks of scientists and movie actors and e.g., the network of boards of directors are also examples of affiliation networks.

An important way of analyzing the structure of connectivity of the bipartite graphs for all of the examples discussed above is consideration of the various functions on the set of connections or *matchings* in a bipartite graph.

1.2 Perfect Matchings

We formally define an (undirected) bipartite graph as follows.

Definition 1.2.1 (Bipartite graph). *The bipartite graph is defined as a triple $G = (V_1, V_2; E)$ where V_1, $|V_1| = m$, and V_2, $|V_2| = n$, are two disjoint sets of vertices with $m \leq n$ and $E = \{(i_k, j_k) : i_k \in V_1, j_k \in V_2; k = 1 \ldots d\}$ is a set of edges. If additionally we associate with E a matrix of weights $\mathbf{x} = [x_{i,j}]$ then we refer to $(G, \mathbf{x})$ as a weighted bipartite graph.*

In particular, when $E = V_1 \times V_2$ we have a complete bipartite graph typically denoted by $K(m, n)$. An example of such graph with $m = 6, n = 7$ is presented in Figure 1.1 below.

In order to analyze the structure of the bipartite graphs one often considers various subsets of its set of edges, often referred to as the *matchings* problem. For instance, in an edge-weighted bipartite graph a maximum weighted matching is defined as a bijective subset of E where the sum of the values of the edges has a maximum value. If a graph is not complete (with weights assumed to be positive) the missing edges are added with zero weight and thus the problem for an incomplete graph reduces to the corresponding problem for a complete graph. Finding a maximum weighted matching in a complete graph is known as the assignment problem and may be solved efficiently within time bounded by a polynomial expression in the number of graph nodes, for instance by

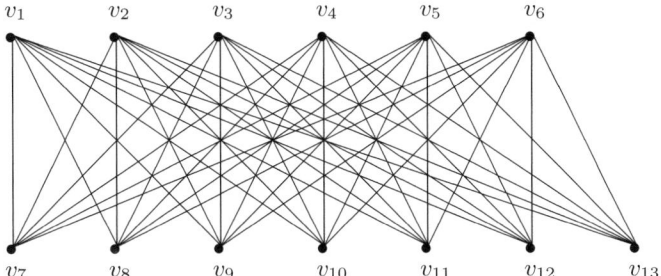

Fig. 1.1. The complete graph K(6, 7).

means of the celebrated *Hungarian algorithm* (Kuhn, 1955; Munkres, 1957). Whereas finding a matching in the optimal assignment problem turns out to be relatively straightforward, some of the more general problems involving finding certain "good" assignments are quite challenging since for very large graphs many of the functions of matchings are very difficult to compute. Indeed, it turns out that often the corresponding counting problems are *#P-complete*, i.e., without known polynomial-time algorithms for their solutions. For instance, a famous result of Valiant (1979) - see also the next section - indicates that counting a number of *perfect matchings* in a bipartite graph is #P-complete. As this notion of a perfect matching shall be crucial to our subsequent discussions, let us define it formally.

Definition 1.2.2 (Perfect matching). *Let* $G = (V_1, V_2; E)$ *be a bipartite graph. Any subgraph* $(V_1, \tilde{V}_2, \tilde{E})$ *of G which represents a bijection between* V_1 *and* $\tilde{V}_2 \subset V_2$ *is called a* perfect matching. *Alternatively, a perfect matching can be identified with an m-tuple* $(j_1, \ldots, j_m)$ *of elements of* V_2 *in the following sense:* $(j_1, \ldots, j_m)$ *is a perfect matching if and only if* $(i, j_i) \in E$ *for* $i = 1, \ldots, m$.

Since the number of perfect matchings is not easy to compute for large graphs, deriving its good approximation is important in many practical special situations. Whenever the graphs (and thus the matchings) of interest are random, one possibility for approximation of such (random) count is by deriving its appropriate limiting distribution and this is indeed one of the main motivations for many of the developments presented in the following chapters of this monograph. In particular, we explore there limiting properties of a wide class of functions of the matchings, namely the class of so called generalized permanent functions. Before formally defining this class we shall first provide some additional background and give examples.

Let $\mathbf{x}$ denote the matrix $[x_{i,j}]$ as in Definition 1.2.1. Let $h : \mathbf{R}^m \to \mathbf{R}$ be a symmetric function (i.e. a function which is invariant under permutations of arguments). Now with any perfect matching considered as an m-tuple $(j_1, \ldots, j_m)$ of elements of V_2 we associate the variables x_{i,j_l}, $l = 1, \ldots, m$, and consequently the value $h(x_{1,j_1}, \ldots, x_{m,j_m})$.

Example 1.2.1 (Binary bipartite graph). For a bipartite graph $G = (V_1, V_2; E)$ let $\mathbf{x}$ be a matrix of binary variables such that $x_{i,j} = 1$ if $(i,j) \in E$, otherwise $x_{i,j} = 0$ and consider a weighted (undirected) bipartite graph $(G, \mathbf{x})$. The matrix $\mathbf{x}$ is often referred to as the (reduced) adjacency matrix of G. Then taking $h(x_1, \ldots, x_m) = x_1 \ldots x_m$ gives

$$h(x_{1,j_1}, \ldots, x_{m,j_m}) = \begin{cases} 1 & \text{when} \ \ x_{1,j_1} = \ldots = x_{m,j_m} = 1 \\ 0 & \text{otherwise.} \end{cases}$$

Thus the function h is simply an indicator of whether a given m-tuple $(j_1, \ldots, j_m)$ is a perfect matching in G.

1.3 Permanent Function

In his famous memoir Cauchy (1812) developed a theory of determinants as special members of a class of alternating-symmetric functions which he distinguished from the ordinary symmetric functions by calling the latter 'permanently symmetric' (*fonctions symétriques permanentes*). He also introduced a certain subclass of the symmetric functions which are nowadays known as 'permanents'.

Definition 1.3.1 (Permanent function of a matrix). *Let* $\mathbb{A} = [a_{i,j}]$ *be a real* $m \times n$ *matrix with* $m \leq n$. *The permanent function of the matrix* $\mathbb{A}$ *is defined by the following formula*

$$Per\,\mathbb{A} = \sum_{(i_1, \ldots, i_m) : \{i_1, \ldots, i_m\} \subset \{1, \ldots, n\}} \prod_{s=1}^{m} a_{s, i_s},$$

where the summation is taken over all possible ordered selections of m *different indices out of the set* $\{1 \ldots, n\}$.

The permanent function has a long history, having been first introduced by Cauchy in 1812 and, almost simultaneously, by Binet (1812). More recently several problems in statistical mechanics, quantum field theory, chemistry as well as enumerative combinatorics and linear algebra have been reduced to the computation of a permanent. Unfortunately, the fastest known algorithm for computing a permanent of an $n \times n$ matrix requires, as shown by Ryser (1963), the exponential in n amount of computational time in the worst-case scenario. Moreover, strong evidence for the apparent complexity of the problem was provided by Valiant (1979) who showed that evaluating a permanent is #P-complete, even when restricted to $0 - 1$ matrices of Example 1.2.1.

In this work we are mostly concerned with *a random permanent* function, meaning that the matrix $\mathbb{A}$ in the above definition is random. Such objects often appear naturally in statistical physics or statistical mechanics problems, when the investigated physical phenomenon is driven by some random process, and thus is stochastic in nature. First we consider the following example relating permanents to matchings in binary bipartite graphs.

Example 1.3.1 (Counting perfect matchings in a bipartite random graph). Let V_1 and V_2 be two sets (of vertices), $|V_1| = m \leq |V_2| = n$. Let $\mathbb{X}$ be a matrix of independent Bernoulli random variables. Define $E = \{(i,j) \in V_1 \times V_2 : \mathbb{X}(i,j) = 1\}$. Then $G = (V_1, V_2; E)$ is a bipartite random graph and $(G, \mathbb{X})$ is the corresponding weighted random graph. Denote the (random) number of *perfect matchings* in $(G, \mathbb{X})$ by $\mathcal{H}(G, \mathbb{X})$. Observe that (cf. also Example 1.2.1)

$$\mathcal{H}(G, \mathbb{X}) = Per\, \mathbb{X}.$$

Our second example is an interesting special case of the first one.

Example 1.3.2 (Network load containment). Consider a communication network modeled by a graph $G(V_1, V_2; E)$ where the each partite set represents one of two types of servers. The network is *operating* if there exists a perfect matching between the two sets of servers. Consider a random graph $(G, \mathbb{X})$ with the matrix of weights $\mathbb{X}$ having all independent and identically distributed entries with the distribution F which corresponds to the assumption that the connections in the network occur independently with distribution F. Hence we interpret the value $X_{i,j}$ as a random load on the connection between server i in set V_1 and server j in set V_2. For a given real value a and a given matching consider a containment function defined as one, if the maximal value of a load in a given matching is less or equals a, and zero otherwise. Let $\mathcal{M}_a(G, \mathbb{X})$ be a total count of such 'contained' matchings in the network. If we define $Y_{i,j} = I(X_{i,j} \leq a)$, where $I(\cdot)$ stands for an indicator function, then

$$\mathcal{M}_a(G, \mathbb{X}) = Per\, \mathbb{Y}.$$

Some basic properties of a permanent function are similar to those of a determinant. For instance, a permanent enjoys a version of the Laplace expansion formula. In order to state the result we need some additional notation.

For any positive integers $k \leq l$ define $\mathcal{S}_{l,k} = \{\mathbf{i} = \{i_1, \ldots, i_k\} : 1 \leq i_1 < \ldots < i_k \leq l\}$ as a collection of all subsets consisting of k elements of a set of l elements. Denote by

$$\mathbb{A}(\mathbf{i}|\mathbf{j}) \quad \mathbf{i} \in \mathcal{S}_{m,r}, \mathbf{j} \in \mathcal{S}_{n,s}$$

the $(r \times s)$ matrix obtained from $\mathbb{A}$ by taking the elements of $\mathbb{A}$ at the intersections of $\mathbf{i} = \{i_1, \ldots, i_r\}$ rows and $\mathbf{j} = \{j_1, \ldots, j_s\}$ columns. Similarly, denote by

$$\mathbb{A}[\mathbf{i}|\mathbf{j}] \quad \mathbf{i} \in \mathcal{S}_{m,r}, \mathbf{j} \in \mathcal{S}_{n,s}$$

the $(m - r) \times (n - s)$ matrix $\mathbb{A}$ with removed rows and columns indexed by $\mathbf{i}$ and $\mathbf{j}$, respectively. The Laplace expansion formula follows now directly from the definition of a permanent (cf., e.g., Minc 1978, chapter 2) and has the following form

$$Per\, \mathbb{A} = \sum_{\mathbf{j} \in \mathcal{S}_{n,s}} Per\, \mathbb{A}(\mathbf{i}|\mathbf{j})\, Per\, \mathbb{A}[\mathbf{i}|\mathbf{j}]$$

for any $\mathbf{i} \in \mathcal{S}_{m,r}$ where $r < m \leq n$. In the "boundary" case $r = m$ we have

$$Per\, \mathbb{A} = \sum_{\mathbf{j} \in \mathcal{S}_{n,m}} Per\, \mathbb{A}(1, \ldots, m|\mathbf{j}). \tag{1.1}$$

The above formula connects a permanent function with a class of symmetric functions known as U-statistics.

1.4 U-statistics

In his seminal paper Hoeffding (1948) extended the notion of an average of l independent and identically distributed (iid) random variables by averaging a symmetric measurable function (kernel) of $k \geq 1$ arguments over $\binom{l}{k}$ possible subsets of k variables chosen out of a set of $l \geq k$ variables. Since then these generalized averages or "U-statistics" have become one of the most intensely studied objects in non-parametric statistics due to its role in the statistical estimation theory.

Let $\mathcal{I}$ be a metric space and denote by $\mathcal{B}(\mathcal{I})$ the corresponding Borel σ-field (i.e., σ-field generated by open sets in $\mathcal{I}$). Let $h : \mathcal{I}^k \to \mathbf{R}$ be a measurable and symmetric (i.e., invariant under any permutation of its arguments) kernel function. Denote the space of all k dimensional kernels h by $L_s^{(k)}$. For $1 \leq k \leq l$ define symmetrization operator $\pi_k^l : L_s^{(k)} \to L_s^{(l)}$ by

$$\pi_k^l\, [h]\, (y_1, \ldots, y_l) = \sum_{1 \leq s_1 < \ldots < s_k \leq l} h(y_{s_1}, \ldots, y_{s_k}) \tag{1.2}$$

for any $(y_1, \ldots, y_l) \in \mathcal{I}^l$.

Definition 1.4.1 (U-statistic). *For any symmetric kernel function $h : \mathcal{I}^k \to \mathbf{R}$ a corresponding U-statistic $U_l^{(k)}(h)$ is given by*

$$U_l^{(k)}(h) = \binom{l}{k}^{-1} \pi_k^l\, [h].$$

In the sequel, whenever it is not ambiguous, we shall write $U_l^{(k)}$ for $U_l^{(k)}(h)$.

In mathematical statistics the arguments of the kernel h (and consequently of $U_l^{(k)}$) are random variables $Y_1, \ldots, Y_l$. Usually it is assumed that they are exchangeable (for instance, equidistributed and independent, see the next section) random elements taking values in a metric space $\mathcal{I}$.

The importance of U-statistics in estimation theory stems from the following fact pointed out by Halmos in his fundamental paper on non-parametric estimation (Halmos, 1946): If θ is a statistical functional defined on a set of probability measures $\mathcal{P}$ on $(\mathcal{I}, \mathcal{B}(\mathcal{I}))$ then it admits and unbiased estimator if and only if there exists a kernel $h \in L_s^{(k)}$ such that

$$\theta(P) = \int \cdots \int h(y_1, \ldots, y_k)\, dP(y_1) \cdots dP(y_k) = Eh(Y_1, \ldots, Y_k), \qquad (1.3)$$

where $Y_1, \ldots, Y_k$ are $\mathcal{I}$-valued, independent identically distributed random variables with the common distribution $P \in \mathcal{P}$. The statistical functionals θ satisfying (1.3) are often called *estimable*. For a given estimable functional θ, the smallest k for which there exists h such that (1.3) holds, is known as its *degree of estimability*.

Suppose now that the family of real-valued distributions $\mathcal{P}$ is large enough to ensure that order statistics $Y_{1:k} \leq \ldots \leq Y_{k:k}$ are *complete*, that is, if f is a symmetric Borel function such that $Ef(Y_1, \ldots, Y_k) = 0$ for all $P \in \mathcal{P}$ then $f = 0$ ($P^{\otimes k}$-a.s. for all $P \in \mathcal{P}$). Then it is easily seen that for a given sample size $l \geq k$ the statistic $U_l^{(k)}(h)$ is a unique ($P^{\otimes l}$-a.s. for all $P \in \mathcal{P}$) unbiased estimator of the statistical functional θ given by (1.3) and thus also the *uniformly minimum variance unbiased* (UMVU) estimator of θ. The condition that ensures that $\mathcal{P}$ is large enough is for instance a requirement that it contains all absolutely continuous distribution functions – an assumption often reasonable in non-parametric statistics. Some examples of U-statistics are given below.

Example 1.4.1 (Examples of U-statistics).

1. *Moments.* If $\mathcal{P}$ is the set of all distributions on the real line with finite mean, then the mean, $\mu = \mu(P) = \int x\, dP(x)$ is an estimable parameter because $f(Y_1) = Y_1$ is an unbiased estimate of μ. The corresponding U-statistic is the sample mean, $U_l^{(1)} = \bar{Y} = (1/l) \sum_{i=1}^{l} Y_i$. Similarly, if $\mathcal{P}$ is the set of all distributions on the real line with finite k-th moment, then the k-th moment, $\mu_k = \int x^k\, dP(x)$ is an estimable parameter with U-statistic, $U_l^{(1)} = (1/l) \sum_{i=1}^{l} Y_i^k$.

2. *Powers of the mean.* For the family $\mathcal{P}$ as above note that for any positive integer k the functional $\mu^k = \mu^k(P)$ is also estimable since $f(Y_1, \ldots, Y_k) = Y_1 \cdots Y_k$ is its unbiased estimator. Thus the corresponding U-statistic is defined by

$$U_l^{(k)} = \binom{l}{k}^{-1} S_l(k)$$

where

$$S_l(k)(y_1, \ldots, y_l) = \pi_k^l[f](y_1, \ldots, y_l) = \sum_{1 \leq s_1 < \ldots < s_k \leq l} y_{s_1} \cdots y_{s_k} \qquad (1.4)$$

is known as an *elementary symmetric polynomial of degree k*.

3. *Variance.* Let $\mathcal{P}$ be the set of all distributions on the real line with finite second moment. The variance $Var\,(P) = \int x^2 dP(x) - \left(\int x dP(x)\right)^2$ is estimable since its unbiased estimator is for instance $f(Y_1, Y_2) = Y_1^2 - Y_1 Y_2$. This estimator is not symmetric however, so we symmetrize it by taking $g(Y_1, Y_2) = (f(Y_1, Y_2) + f(Y_2, Y_1))/2$ which is clearly also unbiased. As a result we obtain the U-statistic based on a two-dimensional kernel

$$s_l^2 = \binom{l}{2}^{-1} \sum_{1 \le s_1 < s_2 \le l} \frac{1}{2}(Y_{s_1} - Y_{s_2})^2 = \frac{1}{l-1}\sum_{i=1}^{l}(Y_i - \bar{Y})^2$$

which is the sample variance.

1.5 The *H*-decomposition

One of the most important tools in investigating U-statistics is the so called *H-decomposition* named after its discoverer W. Hoeffding (Hoeffding 1961). We give a quick overview of H-decomposition below.

For any measure (or signed measure) Q on a measurable space $(\mathcal{I}, \mathcal{B}(\mathcal{I}))$ and a Borel measurable $h : E \to \mathbf{R}$ we denote

$$Q[h] = \int_{\mathcal{I}} h(y)\, Q(dy). \tag{1.5}$$

If Q_i, $1 \le i \le k$, are probability measures (signed measures) on $(\mathcal{I}, \mathcal{B}(\mathcal{I}))$. Then by $\bigotimes_{i=1}^{k} Q_i$ we denote a probability measure (signed measure) on $(\mathcal{I}^k, \mathcal{B}(\mathcal{I}^k))$ which is a product of measures Q_i, $1 \le i \le k$.

In what follows we take $\mathcal{I} = \mathbf{R}$. Let δ_x be a Dirac probability measure at $x \in \mathbf{R}$. Let Y_i, $i = 1, \ldots, k$, be independent and identically distributed real random variables with the law P. Technically the H-decomposition is based on the binomial-like expansion of the right-hand side of the identity

$$h(y_1, \ldots, y_k) = \bigotimes_{i=1}^{k} \delta_{y_i}[h] = \bigotimes_{i=1}^{k} (\delta_{y_i} - P + P)\,[h].$$

More precisely, we will use the following formula valid for any real vectors $(a_1, \ldots, a_m)$ and $(b_1, \ldots, b_m)$

$$\prod_{i=1}^{m}(a_i + b_i) = \prod_{i=1}^{m} b_i + \sum_{r=1}^{m}\sum_{1 \le j_1 < \ldots < j_r \le m} \prod_{i=1}^{r} a_{j_i} \prod_{i=r+1}^{m} b_{j_i} \tag{1.6}$$

where $\{j_1, \ldots, j_r, j_{r+1}, \ldots, j_m\} = \{1, \ldots, m\}$.
Denote

$$h_k(y_1, \ldots, y_k) = h(y_1, \ldots, y_k) = \bigotimes_{i=1}^{k} \delta_{y_i}[h] \tag{1.7}$$

and for $c = 1, \ldots, k - 1$

$$h_c(y_1, \ldots, y_c) = \left(\left(\bigotimes_{i=1}^{c} \delta_{y_i} \right) \otimes P^{\otimes k-c} \right) [h]. \tag{1.8}$$

Note that h_c is the conditional expectation of h with respect to $Y_1, \ldots, Y_c$, i.e.

$$h_c(Y_1, \ldots, Y_c) = E(h|Y_1, \ldots, Y_c)$$

and thus $Eh_c = Eh$ for any $c = 1, \ldots, k$.

Setting in (1.6) $a_i = \delta_{y_i} - P$ and $b_i = P$ and using the fact that kernel h is symmetric we obtain from the representation (1.7) and (1.8) with the help of (1.6) that for $c = 1, \ldots, k$

$$h_c(y_1, \ldots, y_c) - Eh = \sum_{i=1}^{c} \sum_{1 \le j_1 < \ldots < j_i \le c} \left(\bigotimes_{t=1}^{i} (\delta_{y_{j_t}} - P) \right) [h_i]. \tag{1.9}$$

For $i = 1, \ldots, k$, define the *canonical functions* associated with the k-dimensional kernel h by

$$g_i(y_1, \ldots, y_i) = \left(\bigotimes_{j=1}^{i} (\delta_{y_j} - P) \right) [h_i]$$

$$= \int \cdots \int h_i(z_1, \ldots, z_i) \prod_{j=1}^{i} d(\delta_{y_j}(z_j) - P(z_j)).$$

Using again the identity (1.6), this time with $a_i = \delta_{y_i}$, $b_i = -P$ and the representation (1.8) we obtain that

$$g_i(y_1, \ldots, y_i) = \sum_{c=1}^{i} (-1)^{i-c} \pi_c^i [h_c](y_1, \ldots, y_i) + (-1)^i Eh \tag{1.10}$$

or equivalently,

$$g_i(y_1, \ldots, y_i) = \sum_{c=1}^{i} (-1)^{i-c} \pi_c^i [h_c - Eh](y_1, \ldots, y_i) \tag{1.11}$$

Observe that the functions g_i have the property of *complete degeneracy*, that is, for any $1 \le i \le k$ and $j = 1, \ldots, i$

$$Eg_i(y_1, \ldots, Y_j, \ldots, y_i) = Eg_i(y_1, \ldots, y_{i-1}, Y_i)$$

$$= \int g_i(y_1, \ldots, y_{i-1}, z) P(dz) = 0. \tag{1.12}$$

Note that (1.12) implies that under square integrability assumption g_i's are orthogonal in the following sense:

$$E\, g_i(Y_{s_1},\dots,Y_{s_i})g_j(Y_{t_1},\dots,Y_{t_j}) = \begin{cases} E\, g_i^2, & \text{if } \{s_1,\dots,s_i\} = \{t_1,\dots,t_j\}, \\ 0, & \text{if } \{s_1,\dots,s_i\} \neq \{t_1,\dots,t_j\}, \end{cases}$$

where $E\, g_i^2 = E\, g_i^2(Y_1,\dots,Y_i)$.

Rewriting the relationship (1.9) in terms of g_i's we obtain

$$h_c - Eh_c = \sum_{i=1}^{c} \pi_i^c[g_i] \tag{1.13}$$

which is often referred to as the *canonical decomposition* of a symmetric kernel h_c. Taking in the above formula $c = k$ we arrive at the identity for a kernel function on a sequence of equidistributed random variables

$$h(Y_1,\dots,Y_k) - Eh(Y_1,\dots,Y_k) = \sum_{i=1}^{k} \pi_i^k[g_i](Y_1,\dots,Y_k). \tag{1.14}$$

The direct application of the identity (1.14) to the kernel h of a corresponding U-statistic as given by Definition 1.4.1 along with the change of the order of summation gives the *H-decomposition formula* for U-statistics

$$U_l^{(k)} - E\, U_l^{(k)} = \sum_{c=1}^{k} \binom{k}{c} U_{c,l} \tag{1.15}$$

where

$$U_{c,l}(Y_1,\cdots,Y_l) = \binom{l}{c}^{-1} \pi_c^l[g_c](Y_1,\cdots,Y_l)$$

$$= \binom{l}{c}^{-1} \sum_{1 \le i_1 < \dots < i_c \le l} g_c(Y_{i_1},\cdots,Y_{i_c}).$$

Hence it turns out that a U-statistic with given kernel h may be written as a linear combination of U-statistics with corresponding canonical kernels. If for given l, k we have $g_0 = g_1 = \dots = g_{r-1} \equiv 0$ but $g_r \not\equiv 0$, (to make this also work for $r = 1$ we define $g_0 \equiv 0$) then the integer $r - 1 \ge 0$ is called *the level (degree) of degeneration* of $U_l^{(k)}$. Sometimes it is more convenient to work directly with r which is then referred to as a *level of non-degeneracy*. Thus when $r = 1$ then $U_l^{(k)}$ is called *non-degenerate*. The integer $k - r + 1$ is called *the order* of $U_l^{(k)}$. We say that the sequence of U-statistics $(U_l^{(k)})$ is of *infinite order*, if $k - r + 1 \to \infty$ as $l \to \infty$.

The most important features of the H-decomposition are the properties of its components $U_{c,l}$. More precisely, under our assumptions on the $Y's$, for a fixed $c \ge 1$, the following holds true.

(i) $U_{c,l}$ is itself a U-statistic with the kernel function g_c.
(ii) Define a sequence of σ-fields $\mathcal{F}_{c,l} = \sigma\{U_{c,l}, U_{c,l+1}, \ldots\}$, $l = c, c+1, \ldots$. Then

$$E(U_{c,l}|\mathcal{F}_{c,l+1}) = U_{c,l+1}, \quad \forall l = c, c+1, \ldots$$

which implies that the sequence $(U_{c,l}, \mathcal{F}_{c,l})_{l=c,c+1,\ldots}$ is a backward martingale (see the next chapter).
(iii) Additionally, we also have under further integrability assumptions that

$$Cov(U_{c_1,l}, U_{c_2,l}) = 0 \quad \text{for } c_1 \neq c_2 \text{ and } l \geq \max\{c_1, c_2\}.$$

The property (i) follows immediately from the definition of g_c. The property (ii) will be discussed in more detail in the next chapter where we show it to be the consequence of a general result for a larger class of functions (cf. Theorem 2.2.2 (i)). The property (iii) follows from the orthogonality of g_c's. Its more general version will be given in Theorem 2.2.2 (ii).

It is perhaps useful to notice that the property (iii) above implies that a non degenerate U-statistic $U_l^{(k)}$ may be always represented as a sum of identically distributed variables plus an orthogonal reminder. Indeed, assuming the square integrability condition for the kernel h let $\tilde{h} = h - Eh$ and $\tilde{h}_1 = h_1 - Eh_1$ and set $\hat{h}(y_1, \ldots, y_k) = \tilde{h}(y_1, \ldots, y_k) - \sum_{i=1}^{k} \tilde{h}_1(y_i)$. Denote also

$$R_l^{(k)} = \pi_k^l[\hat{h}] = \sum_{c=2}^{k} \binom{k}{c} U_{c,l}.$$

But $U_{1,l}(Y_1, \ldots, Y_l) = \sum_{i=1}^{l} g_1(Y_i)$ and thus for a non-degenerate U-statistic $U_l^{(k)}(h)$, on noting that $\tilde{h}_1 = g_1$ we have

$$U_l^{(k)}(h) - EU_l^{(k)}(h) = \frac{k}{l} \sum_{i=1}^{l} g_1(Y_i) + R_l^{(k)}(Y_1 \ldots, Y_l) \qquad (1.16)$$

where $R_l^{(k)}$ is a U-statistic with kernel a degree of degeneration equal at least one which is orthogonal to $U_{1,l}$, that is,

$$Cov(R_l^{(k)}, U_{1,l}) = 0.$$

Assuming as above that $h(Y_1, \ldots, Y_k)$ is square integrable we may relate the variances of the conditional expectations h_c's, and the canonical kernels g_i's. In particular from (1.13) and (1.12) it follows that for $c = 1, \ldots, k$

$$Var\, h_c = \sum_{i=1}^{c} \binom{c}{i} Var\, g_i. \qquad (1.17)$$

From the above it follows in particular that if r is the level of non-degeneracy, then we have for $c = 1, \ldots, k-1$

$$\frac{Var\, h_c}{\binom{c}{r}} \le \frac{Var\, h_{c+1}}{\binom{c+1}{r}}. \tag{1.18}$$

In order to invert the relationship (1.17) we fix j such that $1 \le j \le k$ and first multiplying both sides of (1.17) by $(-1)^{j-c}\binom{j}{c}$ sum for $c = 1, \ldots, j$.

$$\sum_{c=1}^{j}(-1)^{j-c}\binom{j}{c}Var\, h_c$$

$$= \sum_{c=1}^{j}(-1)^{j-c}\binom{j}{c}\sum_{i=1}^{c}\binom{c}{i}Var\, g_i = \sum_{i=1}^{j}Var\, g_i \sum_{c=i}^{j}(-1)^{j-c}\binom{j}{c}\binom{c}{i}$$

$$= \sum_{i=1}^{j}Var\, g_i \binom{j}{i}\sum_{c=i}^{j}(-1)^{j-c}\binom{j-i}{c-i}$$

$$= Var\, g_j,$$

since

$$\binom{j}{i}\sum_{c=i}^{j}(-1)^{j-c}\binom{j-i}{c-i} = \begin{cases} 1 & i = j, \\ 0 & \text{otherwise.} \end{cases}$$

Thus the dual formula to (1.17) is

$$Var\, g_j = \sum_{c=1}^{j}(-1)^{j-c}\binom{j}{c}Var\, h_c \tag{1.19}$$

for $j = 1, \ldots, k$.

1.6 P-statistics

There exists an obvious, even seemingly trivial, connection between a permanent of a random matrix and a U-statistic with a product kernel as described in Example 1.4.1 part 2. Assume that $Y_1, \ldots, Y_n$ are independent identically distributed real random variables with a finite first moment. Let $\underline{Y}$ denote a so called *one dimensional projection* random matrix. Such a matrix is defined as a $m \times n$ matrix obtained from m row-replicas of the row-vector $(Y_1, \ldots, Y_n)$. Then by the definition of permanent, as given in the Section 1.3,

$$Per\, \underline{Y} = Per \begin{bmatrix} Y_1 & Y_2 & \ldots & Y_n \\ Y_1 & Y_2 & \ldots & Y_n \\ \cdots\cdots\cdots\cdots \\ Y_1 & Y_2 & \ldots & Y_n \end{bmatrix} = m!\, \pi_m^n \left[\prod_{i=1}^{m} Y_i\right]. \tag{1.20}$$

Thus, it can be represented through a U-statistic with the product kernel $h(y_1, \ldots, y_m) = y_1 \cdots y_m$ as

$$Per \, \underline{Y} = \binom{n}{m} m! \, U_n^{(m)}[h] . \tag{1.21}$$

Definition 1.6.1 (Generalized permanent and P-statistic). *For a given kernel function h the generalized permanent function $Per_h \, \mathbb{A}$ is defined as*

$$Per_h \, \mathbb{A} = \sum_{(i_1,\ldots,i_m):\{i_1,\ldots,i_m\}\subset\{1,\ldots,n\}} h(a_{1,j_1}, a_{2,j_2}, \ldots, a_{m,j_m}).$$

(In particular, if $h(x_1, \ldots, x_m) = \prod_{i=1}^{m} x_i$ then $Per_h \, \mathbb{A} = Per\mathbb{A}$). A corresponding P-statistic is then defined as $\binom{n}{m}^{-1} Per_h \, \mathbb{A}/m!$

Note that for a one dimensional projection matrix the above definition reduces to that of a U-statistic from Definition 1.4.1, in view of the relation similar to (1.21) extended now to an arbitrary symmetric kernel function h. Hence, in our present notation, if $\underline{X}$ is a one dimensional projection matrix of m rows and n columns and h is some kernel function of degree m then

$$U_n^{(m)}(h) = \binom{n}{m}^{-1} Per_h \, \underline{X}/m! . \tag{1.22}$$

The formula (1.22) indicates that the theory of U-statistics (perhaps of infinite order) may prove insightful for investigating the behavior of random permanents. However, in order to tackle some of the counting problems for random matrices identified in the introduction we need to extend the relation (1.21) by removing the restrictions on the structure of the matrix as well as on the form of a kernel function. Before proceeding, let us introduce an additional concept which shall be useful in the sequel.

Definition 1.6.2 (Exchangeable random variables). *A random vector $(Y_1, \ldots, Y_n)$ is said to be exchangeable if its joint distribution is invariant under any permutation of the indices $1, \ldots, n$. A family of random variables $(Y_t)_{t\in T}$ is said to be exchangeable if its every finite subset is exchangeable.*

Note that the definition implies that if $(Y_1, \ldots, Y_n)$ is an exchangeable random vector then all Y_i's are identically distributed. It is obvious that a sequence of independent, identically distributed random variables is exchangeable. Below we give a simple example of a dependent exchangeable sequence.

Example 1.6.1 (Dependent exchangeable sequence). Let $(Y_k)_{k\geq 1}$ be a sequence of independent identically distributed real random variables and let Y be such a random variable that Y and the sequence $(Y_k)_{k\geq 1}$ are independent. Define $X_k = Y_k + Y$. Then it is easy to see that that the sequence $(X_k)_{k\geq 1}$ is exchangeable but its elements are not independent.

Exchangeable random variables shall be discussed in greater detail in Chapter 3. In particular, for further discussion of the above example, see Example 3.2.1 therein.

We are returning now to the problem of extending the relationship (1.21). Albeit in what follows this restriction is not necessary, for the sake of simplicity we consider as before the case $\mathcal{I} = \mathbf{R}$.

Let $\mathbb{X}^{(\infty)} = [X_{i,j}]$ be a real random infinite matrix satisfying the following two structural conditions:

(A1) Any column of $\mathbb{X}^{(\infty)}$ is an exchangeable sequence of random variables.
(A2) The columns of $\mathbb{X}^{(\infty)}$ are independent identically distributed random sequences.

Note that the assumptions (A1-A2) imply in particular that all entries of $\mathbb{X}^{(\infty)}$ are identically distributed. These assumptions are also chosen so as to cover the cases of both a one dimensional projection matrix described above and a matrix of independent and identically distributed entries.

By $\mathbb{X}^{(m,n)}$ we denote the submatrix created from the first m rows and n columns of $\mathbb{X}^{(\infty)}$.

1.7 Examples

With the class of generalized permanents defined in the previous section we may now look at some more sophisticated examples of counting problems for matchings in bipartite graphs $G = (V_1, V_2; E)$, $\#(V_1) = m \leq \#(V_2) = n$. Let $\mathbf{x}$ be a matrix of weights and $(G, \mathbf{x})$ the weighted version of G. Consider any symmetric function h of m variables. Then $Per_h(\mathbf{x})$ sums over all perfect matchings in the complete graph $K(m,n)$ the values of the function h calculated for every matching.

To use the concept of the generalized permanent for calculating sums of h values for perfect matchings in any graph $(G, \mathbf{x})$ we define the *edge indicators* as $y_{i,j} = I_E((i,j))$, i.e. $y_{i,j} = 1$ if there is an edge between $i \in V_1$ and $j \in V_2$ and $y_{i,j} = 0$ otherwise. One may think about $[y_{i,j}]$ as an additional set of weights associated with the weighted graph $(G, \mathbf{x})$. Consider a version of the kernel h restricted to perfect matchings in G as

$$\hat{h}((x_1, y_1), \ldots, (x_m, y_m)) = y_1 \ldots y_m h(x_1, \ldots, x_m).$$

Then $\hat{h}((x_{1,j_1}, y_{1,j_1}), \ldots, (x_{m,j_m}, y_{m,j_m}))$ calculates the h-value for $(j_1, \ldots, j_m)$ only if it is a perfect matching in G and otherwise it returns the value zero. (Note that the x-values for the pairs $(i,j) \notin E$ are non-relevant for the value of $\hat{h}$.) In that sense $Per_{\hat{h}}$ is the sum of h-values for all the perfect matchings in $(G, \mathbf{x})$. Obviously, for the complete graph $K(m,n)$ we have $\hat{h} = h$.

If in the above setting $h \equiv 1$ then $Per_{\hat{h}}$ returns the number of perfect matchings in G. Alternatively, this can be achieved by taking h of the product form, i.e. $h(x_1, \ldots, x_m) = x_1 \ldots x_m$ with edge indicators as entries of the $\mathbf{x}$ matrix: $x_{i,j} = y_{i,j}$. Then the number of perfect matchings in $(G, \mathbf{x})$ is just the permanent, $Per_h(\mathbf{x}) = Per(\mathbf{x})$ as discussed in Example 1.3.1.

Now we present some more examples of counting problems for perfect matchings, which are related to generalized permanents. The first one extends Example 1.2.1.

Example 1.7.1 (Multicolored graph). For a complete graph $G = K(m, n)$, consider a weighted graph $(G, \mathbf{x})$ with colored edges, i.e. $x_{i,j} = k$ if the edge (i, j) is painted with the color which is coded by the number k, $k = 1, 2, \ldots, N \leq \infty$. Then the number of perfect matchings which are monochromatic equals $Per_h(\mathbf{x})$ where

$$h(x_1, \ldots, x_m) = \sum_{k=1}^{N} \prod_{i=1}^{m} I(x_i = k).$$

Note that for $N = 2$ this reduces to the count of all perfect matchings as in Examples 1.3.1 and 1.3.2.

The same problem for a graph which is not complete can be treated similarly by regarding some of the edges of the complete graph as being transparent.

Example 1.7.2 (Multicolored graph (cont)). In the setting of the previous example we can add the value 0 for the color codes, meaning that there is no edge (the edge is transparent). Thus $x_{i,j} = 0$ if there is no edge between i and j and it is easy to see that the number of perfect matchings in $(G, \mathbf{x})$ which are monochromatic is just $Per_h(\mathbf{x})$.

Another parameter of interest may be the number of perfect matchings with edges of a given combination of colors.

Example 1.7.3 (Bicolored graph). Consider a complete bipartite graph $K(m, n)$ with edges being painted either red or black. To compute number of matchings with exactly L red edges, first, we associate with each edge a weight $x_{i,j}$ which equals 1 if the edge (i, j) is red and equals 0 if the edge is black. Then by taking

$$h(x_1, \ldots, x_m) = I(x_1 + \ldots + x_m = L)$$

it follows that the number of perfect matchings with L red edges equals $Per_h(\mathbf{x})$.

Our final example in this chapter is concerned with counting heavy and light perfect matchings.

Example 1.7.4 (Heavy and light matchings). Suppose we are interested in matchings which are *heavy* in the sense that the sum of the weights exceeds a given level α. Likewise, we might be interested in counting *light* matchings, that is such that the sum of their weights is below a level β. Then for

$$h_1(x_1,\ldots,x_m) = I(x_1 + \ldots + x_m > \alpha)$$

the generalized permanent Per_{h_1} yields the total number of heavy matchings. Similarly, Per_{h_2} yields the total number of light matchings for

$$h_2(x_1,\ldots,x_m) = I(x_1 + \ldots + x_m < \beta).$$

Alternatively, for a fixed s (usually small) one can be interested in matchings with at most s weights below a given level α or at most s weights above a given level β. In the first case we take

$$h_1(x_1,\ldots,x_m) = I(x_{s+1:m} \geq \alpha), \quad s = 0,1,\ldots,m-1,$$

where $x_{k:n}$ denotes the k-th smallest value among $\{x_1,\ldots,x_n\}$. If $h_1 = 1$ then we say that this represents a matching with few (at most s) light (with weights less than α) edges. $Per_{\hat{h}_1}$ returns the total number of such perfect matchings.

In the second case we take

$$h_2(x_1,\ldots,x_m) = I(x_{m-s:m} < \beta), \quad s = 0,1,\ldots,m-1.$$

If $h_2 = 1$ then we can say that the above represents a perfect matching with a few (at most s) heavy (the weight greater that β). As before, $Per_{\hat{h}_2}$ returns the total count of such perfect matchings.

1.8 Bibliographic Details

Some general references to applications of bipartite graphs in complex systems related to communication and social networks are e.g., in the review articles by Guillaume and Latapy (2004) and Newman et al. (2002). The former article also gives further reference to biological networks applications. In the context of the systems of chemical reactions equations the material in chapter 2 of the recent monograph by Wilkinson (2006) is a good general introduction to biochemical reactions random graphs models (albeit typically in the context of chemical kinetics) and stochastic Petri nets. A good general tutorial on Petri nets is provided by Murata (1989) an the more extensive treatment of the subject is provided in the monograph of Peterson (1981).

The permanent function appeared for the first time in Cauchy's memoirs on determinants (Cauchy, 1812) and, almost simultaneously, in the work by Binet (1812). Until recently for many non-experts the permanent function was associated mainly with an intriguing Van-Der Varden conjecture which was only recently shown to be true. A review article by Bapat (1990) is a good source of information on that topic. The best modern source of results on permanents is the book by Minc (1978). The notion of a random permanent in connection with U-statistic theory is discussed in Korolyuk and Borovskikh (1991), Borovskikh and Korolyuk (1994) and in Rempała and Wesołowski (1999). Ryser's seminal paper on the permanent function computational complexity is also of interest (Ryser, 1963). The proof that for $0 - 1$ matrices evaluating a permanent belongs to $\#P$- class of the computational complexity is given in Valiant (1979). Some more recent results on approximate algorithms for computing permanent are given e.g., in Jerrum et al. (2004). The connections of a permanent function with some statistical problems were reviewed in Bapat (1990).

The classical U-statistics of finite order were introduced in the papers of Halmos (1946) and Hoeffding (1948). The latter paper gave a fundamental result on the asymptotic normality for non-degenerate U-statistics. There are several good general references on U-statistics, e.g., Lee (1990), Koroljuk and Borovskich (1994) or Serfling (1980, chapter 5).

2

Properties of *P*-statistics

2.1 Preliminaries: Martingales

In order to discuss properties of *P*-statistics we first need to review briefly
the basic results on martingales. The theory of martingales is rich and pro-
found and reaches far beyond the basic results stated here. Interested reader
is referred to one of many excellent texts available, like for instance Billingsley
(1999), Chow and Teicher (1978) or Ethier and Kurtz (1986).

For $T \subset \mathbf{R}$ consider an integrable stochastic process $(X_t)_{t \in T}$ defined on a
probability space $(\Omega, \mathcal{F}, P)$.

Definition 2.1.1 (Martingale). *A family of σ-fields $(\mathcal{F}_t)_{t \in T}$ is called a fil-
tration if $\mathcal{F}_t \subset \mathcal{F}$ for any $t \in T$ and $\mathcal{F}_s \subset \mathcal{F}_t$ for any $s \leq t$, $s, t \in T$. The
process $(X_t, \mathcal{F}_t)_{t \in T}$ is called a martingale if*

(a) X_t is $\mathcal{F}_t$-measurable for any $t \in T$
(b) $E(X_t | \mathcal{F}_s) = X_s$ for any $s \leq t$, $s, t \in T$.

It turns out that in the theory of *P*-statistics a related concept of a backward
(or reversed) martingale is also useful.

Definition 2.1.2 (Backward martingale). *Consider a family of σ-fields
$(\mathcal{G}_t)_{t \in T}$ such that $\mathcal{G}_t \subset \mathcal{F}$ for any $t \in T$ and $\mathcal{G}_t \subset \mathcal{G}_s$ for any $s \leq t$, $s, t \in T$.
The process $(X_t, \mathcal{G}_t)_{t \in T}$ is called a backward martingale if*

(a) X_t is $\mathcal{G}_t$-measurable for any $t \in T$
(b) $E(X_s | \mathcal{G}_t) = X_t$ for any $s \leq t$, $s, t \in T$.

Let us prove the following result known as the Doob or maximal inequality
(see, Chow and Teicher 1978, chapter 7 or Ethier and Kurtz 1986, chapter 2).

Theorem 2.1.1 (Maximal inequality for martingales). *Let n be an
arbitrary positive integer and $(X_t, \mathcal{F}_t)_{t \in T}$ be a martingale with a countable
index set T satisfying $\{0, n\} \subset T \subset [0, \infty)$. Then for any $\varepsilon > 0$*

$$P\left(\sup_{t\in T\cap[0,n]}|X_t|>\varepsilon\right)\le\varepsilon^{-2}EX_n^2.$$

Proof. Assume first that $T\cap[0,n]=\{0=t_0,t_1,\ldots,t_M=n\}$ is finite, $t_i<t_{i+1}$, $i=0,1,\ldots,M-1$. For an arbitrary $\varepsilon>0$ define

$$A_k=\cap_{i=0}^{k-1}\{|X_{t_i}|\le\varepsilon\}\cap\{|X_{t_k}|>\varepsilon\},\quad k=0,\ldots,M.$$

Note that A_k, $k=0,\ldots,M$, are disjoint, $A_k\in\mathcal{F}_k$ for each k and

$$\left\{\sup_{t\in T\cap[0,n]}|X_t|>\varepsilon\right\}=\left\{\max_{t\in T\cap[0,n]}|X_t|>\varepsilon\right\}=\bigcup_{k=0}^M A_k.$$

Moreover, due to the martingale property,

$$E[(X_n-X_{t_k})X_{t_k}I(A_k)]=E[X_{t_k}I(A_k)E[X_n-X_{t_k}|\mathcal{F}_{t_k}]]=0.$$

In view of the above, we may write

$$EX_n^2=\sum_{k=0}^M EX_n^2 I(A_k)$$

$$=\sum_{k=0}^M E[(X_n-X_{t_k})^2 I(A_k)]+E[X_{t_k}^2 I(A_k)]$$

$$\ge\sum_{k=0}^M E[X_{t_k}^2 I(A_k)]$$

$$\ge\sum_{k=0}^M \varepsilon^2 P(A_k)=\varepsilon^2 P\left(\bigcup_{k=0}^M A_k\right)$$

$$=\varepsilon^2 P(\sup_{t\in T\cap[0,n]}|X_t|>\varepsilon).$$

In order to prove the result for countable T consider a sequence $T_0\subset T_1\subset T_2\subset\ldots$ of finite subsets of T such that $\{0,n\}\in T_0$ and $\bigcup_k T_k=T$. Then by continuity from below of the probability it follows that for any $\varepsilon>0$

$$P\left(\sup_{t\in T\cap[0,n]}|X_t|>\varepsilon\right)=P\left(\bigcup_k\left\{\max_{t\in T_k\cap[0,n]}|X_t|>\varepsilon\right\}\right)$$

$$=\lim_{k\to\infty}P\left(\max_{t\in T_k\cap[0,n]}|X_t|>\varepsilon\right)\le\varepsilon^{-2}EX_n^2.\qquad\square$$

The result above may be easily extended to the case of an uncountable index set, say, $T=[0,1]$, if we assume some regularity conditions on the trajectories of $(X_t)_{t\in T}$.

Proposition 2.1.1. *Assume that $T = [0, 1]$ and the martingale $(X_t)_{t \in T}$ has trajectories which are a.s. cadlag, i.e., are right continuous and have left-hand limits a.s. Then*

$$P(\sup_{t \in [0,1]} |X_t| > \varepsilon) \leq \varepsilon^{-2} E X_1^2.$$

□

In a similar way one may prove an analogous result for backward martingales. Let us state it for the record as

Theorem 2.1.2 (Maximal inequality for backward martingales). *Let n be an arbitrary positive integer and $(X_t, \mathcal{G}_t)_{t \in T}$ be a backward martingale with $\{n\} \subset T \subset [n, \infty)$ and T being a countable set. Then for any $\varepsilon > 0$*

$$P(\sup_{t \in T \cap [n,\infty)} |X_t| > \varepsilon) \leq \varepsilon^{-2} E X_n^2.$$

Proof. The proof is similar to that of Theorem 2.1.1 and thus omitted.   □

2.2 *H*-decomposition of a *P*-statistic

We first give the result extending the *H*-decomposition (1.15) to an arbitrary *P*-statistic.

Consider an infinite random matrix $\mathbb{X}^{(\infty)}$ satisfying the assumptions (A1-A2) as it was introduced in the previous chapter. We denote by $\mathbb{X}(i_1, \ldots, i_p | j_1, \ldots, j_q)$ a sub-matrix constructed out of $\mathbb{X}^{(m,n)}$ by taking only the entries at the intersections of the selected p rows $\{i_1, \ldots, i_p\} \subseteq \{1, \ldots, m\}$ and q columns $\{j_1, \ldots, j_q\} \subseteq \{1, \ldots, n\}$.

Note that $Per_h \mathbb{X}(i_1, \ldots, i_p | j_1, \ldots, j_q)$ may be regarded as either row- or column- symmetric function. Thus we may apply to it, in both settings, the general symmetrization operation defined by (1.2) with $\mathcal{I} = \mathbf{R}^q$, $k = p, l = m$ when regarded as a row function and with $\mathcal{I} = \mathbf{R}^p$, $k = q, l = n$ when regarded as a column function. Note that under the assumptions (A1-A2) the exchangeability requirement for the random row or column vectors is satisfied.

Denote the row symmetrization of a generalized permanent by

$$Per_h \mathbb{X}(p | j_1, \ldots, j_q) = \sum_{1 \leq i_1 < \ldots < i_p \leq m} Per_h \mathbb{X}(i_1, \ldots, i_p | j_1, \ldots, j_q)$$

$$= \pi_p^m Per_h \mathbb{X}(\cdot | j_1, \ldots, j_q)$$

for any $\{j_1, \ldots, j_q\} \subset \{1, \ldots, n\}$ and the similar column symmetrization by

$$Per_h \mathbb{X}(i_1, \ldots, i_p | q) = \sum_{1 \leq j_1 < \ldots < j_q \leq n} Per_h \mathbb{X}(i_1, \ldots, i_p | j_1, \ldots, j_q)$$

$$= \pi_q^n Per_h \mathbb{X}(i_1, \ldots, i_p | \cdot)$$

for any $\{i_1, \ldots, i_p\} \subset \{1, \ldots, m\}$.

Similarly,

$$
\begin{aligned}
Per_h \, \mathbb{X} \, (p|q) &= \sum_{1 \leq j_1 < \ldots < j_q \leq n} Per_h \, \mathbb{X} \, (p|j_1, \ldots, j_q) \\
&= \sum_{1 \leq i_1 < \ldots < i_p \leq m} Per_h \, \mathbb{X} \, (i_1, \ldots, i_p|q) \\
&= \sum_{1 \leq j_1 < \ldots < j_q \leq n} \sum_{1 \leq i_1 < \ldots < i_p \leq m} Per_h \, \mathbb{X} \, (i_1, \ldots, i_p|j_1, \ldots, j_q).
\end{aligned}
$$

Using this notation, we may now easily extend the Laplace expansion formula for permanents given by (1.1). Let us state it in the form of the following lemma.

Lemma 2.2.1. *Let* $\mathbb{X}^{(m,n)}$ *be an* $m \times n$-*matrix with* $1 \leq m \leq n$. *Then*

$$
Per_h \, \mathbb{X}^{(m,n)} = Per_h \, \mathbb{X} \, (1, 2, \ldots, m|m). \tag{2.1}
$$

$\square$

The above formula suggests that in order to decompose $Per_h \, \mathbb{X}$, it suffices to decompose a generalized permanent of a square submatrix with columns $j_1 \ldots, j_m$

$$
Per_h \, \mathbb{X} \, (1, 2, \ldots, m|j_1, j_2, \ldots, j_m).
$$

Assume for convenience that $Eh = 0$ (otherwise taking $\tilde{h} = h - Eh$ in what follows).

Denote by S_q a set of permutations of $\{1, \ldots, q\}$. In view of the above (cf. (1.14)) the decomposition of $Per_h \, \mathbb{X} \, (1, 2, \ldots, m|j_1, j_2, \ldots, j_m)$ can be obtained as follows

$$
\begin{aligned}
& Per_h \, \mathbb{X} \, (1, 2, \ldots, m|j_1 \ldots, j_m) \\
&= \sum_{\sigma \in S_m} h(X_{1,j_{\sigma(1)}}, X_{2,j_{\sigma(2)}}, \ldots, X_{m,j_{\sigma(m)}}) \\
&= \sum_{\sigma \in S_m} \sum_{c=1}^{m} (\pi_c^m g_c)(X_{1,j_{\sigma(1)}}, X_{2,j_{\sigma(2)}}, \ldots, X_{m,j_{\sigma(m)}}) \\
&= \sum_{c=1}^{m} \sum_{1 \leq i_1 < \ldots < i_c \leq m} \sum_{\sigma \in S_m} g_c(X_{i_1,j_{\sigma(i_1)}}, \ldots, X_{i_c,j_{\sigma(i_c)}}).
\end{aligned}
$$

But by the definition of a generalized permanent and in view of (2.1)

$$\sum_{\sigma \in S_m} g_c(X_{i_1,j_{\sigma(i_1)}}, \ldots, X_{i_c,j_{\sigma(i_c)}})$$

$$= \sum_{1 \le l_1 < \ldots < l_c \le m} \sum_{\tau \in S_c} g_c(X_{i_{\tau(1)},j_{l_1}}, \ldots, X_{i_{\tau(c)},j_{l_c}})$$

$$= \sum_{1 \le l_1 < \ldots < l_c \le m} Per_{g_c} \mathbb{X}(i_1, \ldots, i_c | j_{l_1}, \ldots, j_{l_c})$$

$$= Per_{g_c} \mathbb{X}(i_1, \ldots, i_c | j_1, \ldots, j_m).$$

Hence

$$Per_h \mathbb{X}(1, 2, \ldots, m | j_1 \ldots, j_m)$$

$$= \sum_{c=1}^{m} \sum_{1 \le i_1 < \ldots < i_c \le m} Per_{g_c} \mathbb{X}(i_1, \ldots, i_c | j_1, \ldots, j_m)$$

$$= \sum_{c=1}^{m} Per_{g_c} \mathbb{X}(c | j_1, \ldots, j_m).$$

Thus, using again (2.1), we may write

$$Per_h \mathbb{X}^{(m,n)} = Per_h \mathbb{X}(1, 2, \ldots, m | m)$$

$$= \sum_{1 \le j_1 < \ldots < j_m \le n} Per_h \mathbb{X}(1, 2, \ldots, m | j_1, \ldots, j_m)$$

$$= \sum_{1 \le j_1 < \ldots < j_m \le n} \sum_{c=1}^{m} Per_{g_c} \mathbb{X}(c | j_1, \ldots, j_m)$$

$$= \sum_{c=1}^{m} \binom{n-c}{m-c} (m-c)! \sum_{1 \le j_1 < \ldots < j_c \le n} Per_{g_c} \mathbb{X}(c | j_1, \ldots, j_c)$$

$$= \sum_{c=1}^{m} \binom{n-c}{m-c} (m-c)! \, Per_{g_c} \mathbb{X}(c | c).$$

Rewriting the above as

$$Per_h \mathbb{X}^{(m,n)} = \sum_{c=1}^{m} m! \binom{n}{m} \binom{n}{c}^{-1} Per_{g_c} \mathbb{X}(c | c)/c!$$

and denoting

$$U_{g_c}^{(m,n)} = \binom{m}{c}^{-1} \binom{n}{c}^{-1} Per_{g_c} \mathbb{X}(c | c)/c! \qquad (2.2)$$

we may summarize the above considerations as follows.

Theorem 2.2.1. *Under assumptions (A1-A2) on the entries of a matrix* $\mathbb{X}^{(m,n)}$, *suppose also that the kernel h satisfies* $E|h| < \infty$. *Then*

$$\binom{n}{m}^{-1} \left(Per_h \, \mathbb{X}^{(m,n)} - EPer_h \, \mathbb{X}^{(m,n)} \right) / m! = \sum_{c=1}^{m} \binom{m}{c} U_{g_c}^{(m,n)},$$

where the components $U_{g_c}^{(m,n)}$ are defined by (2.2).   □

If $\mathbb{X}^{(m,n)} = \underline{X}$ that is, $\mathbb{X}^{(m,n)}$ is a one dimensional projection matrix then by (1.22) a P-statistic is a U-statistic and the Theorem 2.2.1 reduces to the H-decomposition (1.15). Let us also note that for any fixed $c \geq 1$ the component $U_{g_c}^{(m,n)}$ is simply a P-statistic obtained by symmetrization (with respect to rows and columns of $\mathbb{X}^{(m,n)}$) of a generalized permanent with a kernel function g_c based on a square $c \times c$ submatrix of $\mathbb{X}^{(m,n)}$.

For the purpose of illustration let us provide a simple example of an application of Theorem 2.2.1.

Example 2.2.1 (Random permanent decomposition). Let $E \, X_{i,j} = \mu \neq 0$ and consider the kernel $h(y_1, \ldots, y_m) = \prod_{i=1}^{m} (y_i/\mu)$. Then

$$g_c(y_1, \ldots, y_c) = \prod_{i=1}^{c} (y_i - \mu) \, \mu^{-c}$$

and by Theorem 2.2.1 we have

$$\frac{Per \, \mathbb{X}^{(m,n)}}{\binom{n}{m} m! \mu^m} = 1 + \sum_{c=1}^{m} \binom{m}{c} U_c^{(m,n)},$$

where

$$U_c^{(m,n)} = \binom{n}{c}^{-1} \binom{m}{c}^{-1} (\mu^c c!)^{-1} Per \, \tilde{\mathbb{X}} (c|c) \qquad (2.3)$$

$$= \binom{n}{c}^{-1} \binom{m}{c}^{-1} (\mu^c c!)^{-1} \sum_{1 \leq i_1 < \ldots < i_c \leq m}$$

$$\sum_{1 \leq j_1 < \ldots < j_c \leq n} Per \, \tilde{\mathbb{X}} (i_1, \ldots, i_c | j_1, \ldots, j_c), \qquad (2.4)$$

for $\tilde{X}_{i,j} = X_{i,j} - \mu$, $(i = 1, \ldots, m, \ j = 1, \ldots, n)$.

The fact that the generalized H-decomposition given in Theorem 2.2.1 retains the uncorrelated martingale structure of the components $U_{g_c}^{(m,n)}$'s is verified in the following result.

Theorem 2.2.2. *Assume $m = m_n$ is a non-decreasing sequence in n. Under assumptions (A1-A2) on the entries of matrix $\mathbb{X}^{(\infty)}$, for a fixed $c \geq 1$*

(i) if $\mathcal{F}_c^{(n)} = \sigma\{U_{g_c}^{(m_n,n)}, U_{g_c}^{(m_{n+1},n+1)}, \ldots\}$, then for $n_0(c) = \min\{n : m_n \geq c\}$

$$E(U_{g_c}^{(m_n,n)} | \mathcal{F}_c^{(n+1)}) = U_{g_c}^{(m_{n+1},n+1)}, \quad \forall n = n_0(c), n_0(c) + 1, \ldots$$

that is, $\left(U_{g_c}^{(m_n,n)}, \mathcal{F}_c^{(n)} \right)_{n=n_0(c), n_0(c)+1, \ldots}$ *is a backward martingale,*

(ii) the elements $(U_{g_c}^{(m,n)})_c$ of the H-decomposition of a P-statistic are orthogonal, i.e.

$$Cov(U_{g_{c_1}}^{(m,n)}, U_{g_{c_2}}^{(m,n)}) = 0 \qquad \text{if } c_1 \neq c_2.$$

Proof. (i) Let us denote by $U_{g_c}^{(m_n,n)}(l_1, \ldots, l_{m_{n+1}-m_n}; k)$ an element of the Hoeffding decomposition (as given in Theorem 2.2.1) of an $m_n \times n$ matrix, which is obtained by deleting the rows numbered $l_1, \ldots, l_{m_{n+1}-m_n}$ and the k-th column from the $m_{n+1} \times (n+1)$ matrix, i.e.,

$$U_{g_c}^{(m_n,n)}(l_1, \ldots, l_{m_{n+1}-m_n}; k) = \binom{m_n}{c}^{-1} \binom{n}{c}^{-1} (c!)^{-1}$$

$$\times \sum_{\substack{1 \leq i_1 < \ldots < i_c \leq m_{n+1} \\ \{i_1, \ldots, i_c\} \cap \{l_1, \ldots, l_{m_{n+1}-m_n}\} = \emptyset}} \sum_{\substack{1 \leq j_1 < \ldots j_c \leq n+1 \\ k \notin \{j_1, \ldots, j_c\}}} Per_{g_c} \mathbb{X}(i_1, \ldots, i_c | j_1, \ldots, j_c).$$

Observe that

$$\sum_{k=1}^{n+1} \sum_{1 \leq l_1 < \ldots < l_{m_{n+1}-m_n} \leq m_{n+1}} U_{g_c}^{(m_n,n)}(l_1, \ldots, l_{m_{n+1}-m_n}; k)$$

$$= \binom{m_{n+1}-c}{m_n-c} \frac{n+1-c}{\binom{m_n}{c}\binom{n}{c}} Per_{g_c} \mathbb{X}^{(m_{n+1},n+1)}(c|c)/c!,$$

where $\mathbb{X}^{(m_{n+1},n+1)}$ is the $m_{n+1} \times (n+1)$ dimensional matrix created from the left upper corner of the infinite matrix $\mathbb{X}^{(\infty)}$. Since

$$\binom{m_{n+1}-c}{m_{n+1}-m_n} \binom{m_n}{c}^{-1} = \binom{m_{n+1}}{m_n} \binom{m_{n+1}}{c}^{-1},$$

the above entails

$$\sum_{k=1}^{n+1} \sum_{1 \leq l_1 < \ldots < l_{m_{n+1}-m_n} \leq m_{n+1}} U_{g_c}^{(m_n,n)}(l_1, \ldots, l_{m_{n+1}-m_n}; k)$$

$$= (n+1) \binom{m_{n+1}}{m_n} U_{g_c}^{(m_{n+1},n+1)}. \tag{2.5}$$

Let us note that in view of the assumptions (A1-A2) it follows that the conditional distribution of $U_{g_c}^{(m_n,n)}(l_1, \ldots, l_{m_{n+1}-m_n}; k)$ given $\mathcal{F}_c^{(n+1)}$ is the same for any particular choice of $k \in \{1, \ldots, n+1\}$ and $\{l_1, \ldots, l_{m_{n+1}-m_n}\} \subset \{1, \ldots, m_{n+1}\}$. Consequently,

$$E[U_{g_c}^{(m_n,n)}(l_1, \ldots, l_{m_{n+1}-m_n}; k) | \mathcal{F}_c^{(n+1)}]$$
$$= E[U_{g_c}^{(m_n,n)}(n+1, n+2, \ldots, m_{n+1}; n+1) | \mathcal{F}_c^{(n+1)}] = E[U_{g_c}^{(m,n)} | \mathcal{F}_c^{(n+1)}]$$

yielding

$$E\left[\sum_{k=1}^{n+1}\sum_{1\le l_1<\ldots<l_{m_{n+1}-m_n}\le m_{n+1}} U_{g_c}^{(m_n,n)}(l_1,\ldots,l_{m_{n+1}-m_n};k)\middle| \mathcal{F}_c^{(n+1)}\right]$$

(2.6)

$$= (n+1)\binom{m_{n+1}}{m_n} E(U_{g_c}^{(m_n,n)}|\mathcal{F}_c^{(n+1)}).$$

(2.7)

On the other hand, in view of the identity (2.5), we have that (2.6) is equal to

$$(n+1)\binom{m_{n+1}}{m_n} E(U_{g_c}^{(m_{n+1},n+1)}|\mathcal{F}_c^{(n+1)}) = (n+1)\binom{m_{n+1}}{m_n} U_{g_c}^{(m_{n+1},n+1)}. \quad (2.8)$$

Comparing the expression (2.7) with that at the right-hand side of (2.8), we arrive at *(i)*.

In order to prove *(ii)* it is enough to show that for $c_1 \ne c_2$ we have

$$Cov\left(Per_{g_{c_1}}\mathbb{X}(c_1|c_1), Per_{g_{c_2}}\mathbb{X}(c_2|c_2)\right) = 0.$$

To this end we will show that for any pairs of fixed sets of rows $\{i_1,\ldots,i_{c_1}\}$, $\{k_1,\ldots,k_{c_2}\}$ and columns $\{j_1,\ldots,j_{c_1}\}$, $\{l_1,\ldots,l_{c_2}\}$

$$Cov\left[Per_{g_{c_1}}\mathbb{X}(i_1,\ldots,i_{c_1}|j_1,\ldots,j_{c_1}), Per_{g_{c_2}}\mathbb{X}(k_1,\ldots,k_{c_2}|l_1,\ldots,l_{c_2})\right] = 0.$$

(2.9)

Consider an arbitrary pair of fixed sets of rows $\{i_1,\ldots,i_{c_1}\}$, $\{k_1,\ldots,k_{c_2}\}$ and columns $\{j_1,\ldots,j_{c_1}\}$, $\{l_1,\ldots,l_{c_2}\}$. Let us note that by linearity of the covariance operator

$$Cov\left[Per_{g_{c_1}}\mathbb{X}(i_1,\ldots,i_{c_1}|j_1,\ldots,j_{c_1}), Per_{g_{c_2}}\mathbb{X}(k_1,\ldots,k_{c_2}|l_1,\ldots,l_{c_2})\right]$$

$$= \sum_{\sigma\in S_{c_1},\ \tau\in S_{c_2}} E\left(g_{c_1}(X_{i_1,j_{\sigma(1)}},\ldots,X_{i_{c_1},j_{\sigma(c_1)}})\, g_{c_2}(X_{k_1,l_{\tau(1)}},\ldots,X_{k_{c_2},l_{\tau(c_2)}})\right).$$

Since $c_1 \ne c_2$ we may assume without loosing generality that $c_1 > c_2$. It follows that there exist at least one column $j_s \in \{j_1,\ldots,j_{c_1}\}$ such that $j_s \notin \{l_1,\ldots,l_{c_2}\}$. But by the assumption of the independence of columns and the property of canonical functions (1.12) the standard conditioning argument implies now that all the summands above are zero and thus (2.9) follows. The proof of the result is complete. $\qquad\square$

2.3 Variance Formula for a P-statistic

Using the representation of Theorem 2.2.1 it is possible to derive a general formula for the variance of a P-statistic which proves very useful later on.

Theorem 2.3.1 (The variance formula for a P-statistic). *Under the assumptions (A1-A2) on $\mathbb{X}^{(\infty)}$ let us suppose $Eh^2 < \infty$. Then for $c = 1, \ldots, m$*

$$Var\, U_c^{(m,n)} = \binom{n}{c}^{-1}\binom{m}{c}^{-1}\sum_{d=0}^{c}\binom{m-d}{c-d}\frac{D(c,d)}{d!}$$

where

$$D(c,d) = \sum_{s=0}^{d}\binom{d}{s}(-1)^{d-s}\rho_{c,s}$$

for $d = 0, 1, \ldots, c$ and

$$\rho_{c,s} = E\left[g_c(X_{11}, \ldots, X_{ss}, X_{i_{s+1},s+1} \ldots, X_{i_c,c}) \times \right.$$
$$\left. g_c(X_{11}, \ldots, X_{ss}, X_{j_{s+1},s+1} \cdots, X_{j_c,c})\right]$$

where $i_k \neq j_k$ for $k = s + 1, \ldots, c$ and $s = 0, \ldots, c$.

Proof. First, let us note that by Theorem 2.2.2 it follows that

$$Var\,\frac{Per_h\,\mathbb{X}}{\binom{n}{m}m!} = \sum_{c=1}^{m}\binom{m}{c}^2 Var\, U_{g_c}^{(m,n)} + \sum_{1\leq c_1 \neq c_2 \leq m}\binom{m}{c_1}\binom{m}{c_2}$$

$$Cov\left[U_{g_{c_1}}^{(m,n)}, U_{g_{c_2}}^{(m,n)}\right] = \sum_{c=1}^{m}\binom{m}{c}^2 Var\, U_{g_c}^{(m,n)}. \qquad (2.10)$$

Now, for $c = 1, \ldots m$

$$\binom{n}{c}^2\binom{m}{c}^2 c!^2\, Var\, U_{g_c}^{(m,n)} = Var\, Per_{g_c}\,\mathbb{X}\,(c|c)$$

$$= \sum_{1\leq j_1 < \ldots < j_c \leq n} Var\,(Per_{g_c}\,\mathbb{X}\,(c|j_1, \ldots, j_c))$$

since, by independence of columns of $\mathbb{X}^{(\infty)}$, we have (cf. also the proof of part (ii) of the Theorem 2.2.2)

$$Cov\left[Per_{g_c}\,\mathbb{X}\,(c|j_1, \ldots, j_c), Per_{g_c}\,\mathbb{X}\,(c|l_1, \ldots, l_c)\right] = 0$$

if only $\{j_1, \ldots, j_c\} \neq \{l_1, \ldots, l_c\}$.
Since the columns of $\mathbb{X}^{(\infty)}$ are identically distributed, we obtain

$$\binom{n}{c}\binom{m}{c}^2 c!^2\, Var\, U_{g_c}^{(m,n)} = Var\,(Per_{g_c}\,\mathbb{X}\,(c|1, \ldots, c)).$$

Let us note that the number of pairs of $c \times c$ submatrices of $\mathbb{X}^{(\infty)}$ having exactly k rows in common equals $\binom{m}{c}\binom{c}{k}\binom{m-c}{c-k}$, for $\max(0, 2c - m) \leq k \leq c$, and each such pair has equal covariance (since the row vectors of $\mathbb{X}^{(\infty)}$ are identically distributed). Hence, for given c, the above right-hand side can be written as

$$
\binom{m}{c} \sum_{k=\max(0,2c-m)}^{c} \binom{c}{k}\binom{m-c}{c-k}
$$

$$
\times Cov\left[Per_{g_c} \mathbb{X}(1,\ldots,k,i_{k+1},\ldots,i_c|1,\ldots,c),\right.
$$

$$
\left. Per_{g_c} \mathbb{X}(1,\ldots,k,l_{k+1},\ldots,l_c|1,\ldots,c)\right]
$$

where $\{i_{k+1},\ldots,i_c\} \cap \{l_{k+1},\ldots,l_c\} = \emptyset$.

Observe that each term of the above sum is itself a sum of the partial covariance elements of the form

$$
E\left[g_c(X_{i_1,1},\ldots,X_{i_l,l},X_{i_{l+1},l+1}\ldots,X_{i_c,c})\right.
$$

$$
\left. g_c(X_{i_1,1},\ldots,X_{i_l,l},X_{j_{l+1},l+1}\ldots,X_{j_c,c})\right]
$$

where $i_{l+1},\ldots,i_c$ and $j_{l+1},\ldots,j_c$ are fixed non-overlapping subsets of $\{l+1,\ldots,m\}$ for some $0 \le l \le k$. By the assumptions about the entries of the matrix $\mathbb{X}^{(\infty)}$ it follows that the partial covariances having exactly l $(0 \le l \le k)$ elements in common are the same and equal to

$$
\rho_{c,l} = E\left[g_c(X_{1,1},\ldots,X_{l,l},X_{i_{l+1},l+1}\ldots,X_{i_c,c})\right.
$$

$$
\left. g_c(X_{1,1},\ldots,X_{l,l},X_{j_{l+1},l+1}\ldots,X_{j_c,c})\right]
$$

Now, to compute the covariance of such $k \times c$ permanents it suffices to find the number of pairs of c-tuples of arguments of the function g_c with exactly l elements in common, $(0 \le l \le k \le c)$. Observe that it equals to the number of pairs of c-tuples having exactly l common elements in a permanent of the matrix $k \times c$, multiplied by $(c-k)!^2$ – the number of all possible permutations of $i_{k+1},\ldots,i_c$ and $l_{k+1},\ldots,l_c$.

To compute the number of pairs of c-tuples with exactly l elements in common let us start with finding the number of c-tuples present in the defining formula for $Per_{g_c} \mathbb{Y}[k,c]$, where $\mathbb{Y}[k,c]$ is a $k \times c$ matrix, having exactly l elements in common with the diagonal entries $y_{1,1},\ldots,y_{k,k}$. First, we fix l factors in $\binom{k}{l}$ ways. If we assume that $y_{1,1},\ldots,y_{l,l}$ are fixed, then the remaining factors, in the c-tuples we are looking for, have to be of the form $y_{l+1,j_{l+1}},\ldots,y_{k,j_k}$, where $j_d \ne d$, $d = l+1,\ldots,k$. Finding the number of such c-tuples (say, $\mathcal{R}_l(k,c)$) is equivalent to computing the number of summands in a permanent of the matrix of dimensions $(k-l) \times (c-l)$ which do not contain any diagonal entry. To this end, we subtract the number of all summands having at least one factor being the diagonal entry, from the total number of all summands in that permanent. Using the exclusion-inclusion formula we get that

$$\mathcal{R}_l(k,c) = \binom{c-l}{k-l}(k-l)! - \sum_{j=1}^{k-l}(-1)^{j+1}\binom{k-l}{j}\binom{c-l-j}{k-l-j}(k-l-j)!,$$

where the absolute value of the j-th member of the above sum denotes the number of c-tuples having exactly j factors being the diagonal entries (equal to the number of choices of j positions on the diagonal) multiplied by the number of c-tuples of $k-l-j$ factors from the outside of the diagonal (equal to number of c-tuples in the permanent of the matrix of dimensions $(k-l-j) \times (c-l-j)$). Thus, in a slightly more compact form,

$$\mathcal{R}_l(k,c) = \sum_{j=0}^{k-l}(-1)^j\binom{k-l}{j}\binom{c-l-j}{k-l-j}(k-l-j)!. \qquad (2.11)$$

Consequently, the number of pairs of c-tuples in $Per\,\mathbb{Y}[k,c]$ with exactly l factors in common equals to

$$\binom{c}{k}k!\binom{k}{l}\mathcal{R}_l(k,c).$$

Hence, combining the above formula with an earlier one for the number of pairs of c-tuples with l identical factors we arrive at

$$Cov\,[Per_{g_c}\,\mathbb{X}\,(1,\dots,k,i_{k+1},\dots,i_c|1,\dots,c),$$
$$Per_{g_c}\,\mathbb{X}\,(1,\dots,k,l_{k+1},\dots,l_c|1,\dots,c)]$$
$$= (c-k)!^2 \sum_{l=0}^{k}\binom{c}{k}k!\binom{k}{l}\rho_{c,l}\mathcal{R}_l(k,c)\,.$$

Now, returning to the formula for the variance of $U_{g_c}^{(m,n)}$ we obtain by (2.11)

$$\binom{n}{c}\binom{m}{c}Var\,U_{g_c}^{(m,n)}$$

$$= \frac{1}{c!^2}\sum_{k=\max(0,2c-m)}^{c}\binom{c}{k}^2\binom{m-c}{c-k}k!(c-k)!^2\sum_{l=0}^{k}\binom{k}{l}\rho_{c,l}\,\mathcal{R}_l(k,c)$$

$$= \sum_{k=\max(0,2c-m)}^{c}\binom{m-c}{c-k}\sum_{d=0}^{k}\binom{c-d}{k-d}\frac{1}{d!}D(c,d), \qquad (2.12)$$

since

$$\sum_{l=0}^{k}\binom{k}{l}\rho_{c,l}\mathcal{R}_l(k,c) = \sum_{l=0}^{k}\binom{k}{l}\rho_{c,l}\sum_{d=l}^{k}(-1)^{d-l}\binom{k-l}{d-l}\binom{c-d}{k-d}(k-d)!$$

$$= \sum_{d=0}^{k}\binom{c-d}{k-d}(k-d)!\binom{k}{d}\sum_{l=0}^{d}\binom{d}{l}(-1)^{d-l}\rho_{c,l}$$

$$= \sum_{d=0}^{k} \binom{c-d}{k-d} \frac{k!}{d!} D(c,d).$$

Observe that we can further simplify the expression (2.12), since

$$\sum_{k=\max(0,2c-m)}^{c} \binom{m-c}{c-k} \sum_{d=0}^{k} \binom{c-d}{k-d} \frac{D(c,d)}{d!}$$

$$= \sum_{d=0}^{c} \frac{D(c,d)}{d!} \sum_{k=\max(d,2c-m)}^{c} \binom{m-c}{c-k}\binom{c-d}{k-d} = \sum_{d=0}^{c} \binom{m-d}{c-d} \frac{D(c,d)}{d!},$$

where the last equality follows by applying the hypergeometric summation rule for the inner sum. Thus, we can rewrite (2.12) as

$$Var\, U_{g_c}^{(m,n)} = \binom{n}{c}^{-1}\binom{m}{c}^{-1} \sum_{d=0}^{c} \binom{m-d}{c-d} \frac{D(c,d)}{d!},$$

which along with (2.10) completes the proof. $\square$

Example 2.3.1 (Variance of a random permanent). Consider again the kernel function as in Example 2.2.1. Let as before $\mu \neq 0$ be the common mean of $X_{i,j}$ and additionally let $0 < \sigma^2 < \infty$ be the common variance. By $\gamma = \sigma/\mu$ we denote the coefficient of variation and by $\rho > 0$ the correlation between any two components from the same column. Then,

$$\rho_{c,s} = \gamma^{2c}\rho^{c-s}$$

and thus

$$D(c,d) = \gamma^{2c}\rho^{c-d}(1-\rho)^d.$$

As shown before in Example 2.2.1 the canonical function $U_{g_c}^{(m,n)}$ of the H-decomposition is given by (2.3). From Theorem 2.3.1 we have now that

$$Var\, U_c^{(m,n)} = \binom{n}{c}^{-1}\binom{m}{c}^{-1} \gamma^{2c} \sum_{d=0}^{c} \binom{m-d}{c-d} \frac{\rho^{c-d}(1-\rho)^d}{d!},$$

and hence by the orthogonality of $U_{g_c}^{(m,n)}$ for $c = 1, 2, \ldots, m$, it follows that

$$Var\left(\frac{Per\, \mathbb{X}^{(m,n)}}{\binom{n}{m} m!\, \mu^m}\right) = \sum_{c=1}^{m} \frac{\binom{m}{c}\gamma^{2c}}{\binom{n}{c}} \sum_{d=0}^{c} \frac{1}{d!}\binom{m-d}{c-d}(1-\rho)^d\rho^{c-d}.$$

An important special case of Theorem 2.3.1 occurs when the rows of $\mathbb{X}^{(\infty)}$ are also independent. Then the matrix $\mathbb{X}^{(\infty)}$ has all independent and identically distributed entries and

$$D(c,d) = 0 \quad \text{for} \quad d < c.$$

Since $\rho(c, c) = Var\, g_c$, we see that for each canonical component $U_{g_c}^{(m,n)}$ we must have

$$Var\left(U_{g_c}^{(m,n)}\right) = \binom{m}{c}^{-1}\binom{n}{c}^{-1}\frac{Var\, g_c}{c!},$$

which entails (by orthogonality) that the formula for the variance of a P-statistic degenerate of degree $r - 1$ in case when $\mathbb{X}^{(\infty)}$ has all independent and identically distributed entries is

$$Var\left(\frac{Per_h\, \mathbb{X}^{(m,n)}}{\binom{n}{m}m!}\right) = \sum_{c=r}^{m}\frac{\binom{m}{c}}{\binom{n}{c}}\frac{Var\, g_c}{c!}. \tag{2.13}$$

Another important special case is that of a U-statistic which corresponds to $\mathbb{X}^{(\infty)} = \underline{X}$ i.e., the one dimensional projection matrix. Then in Theorem 2.3.1 we have

$$D(c, s) = 0 \quad \text{for} \quad s > 0$$

and obtain the following variance formula for a U-statistic of degree of degeneration $r - 1$ as defined in Definition 1.4.1 with $l = n$ and $k = m$

$$Var\left(\frac{Per_h\, \underline{X}}{\binom{n}{m}m!}\right) = Var\left(U_n^{(m)}(h)\right) = \sum_{c=r}^{m}\frac{\binom{m}{c}^2}{\binom{n}{c}}Var\, g_c. \tag{2.14}$$

We conclude this chapter with the following example illustrating one possible use of Theorem 2.3.1. The example could be considered as a warm-up before the material of the next chapter. Recall that for a sequence of random variables X_n and a sequence of real numbers a_n the notation $X_n = o_p(a_n)$ stands for $\forall_{\varepsilon>0}P(|X_n| > a_n\varepsilon) \to 0$ as $n \to \infty$. In this notation we may state the following simple approximation theorem for non-degenerate P-statistics.

Example 2.3.2 (Approximation theorem for P-statistics). Assume (A1-A2) for matrix $\mathbb{X}^{(\infty)}$. Suppose $m/\sqrt{n} \to 0$ as $n \to \infty$, $Eh^2 < \infty$ and $Var\,(g_1) \neq 0$ where g_1 does not depend on m, n, as well as for some non-random positive constants M_1 and M_2 which do not depend upon m, n, we have for $c = 1, 2, 3, \ldots$

$$\forall_{d\leq c}\, D(c, d) \leq M_1^c M_2^d < \infty.$$

Then

$$\frac{Per_h\, \mathbb{X}^{(m,n)}}{\binom{n}{m}m!} - Eh = \frac{1}{n}\sum_{ij}g_1(X_{ij}) + o_p\left(\frac{m}{\sqrt{n}}\right). \tag{2.15}$$

The formula (2.15) is a simple consequence of the properties of P-statistics summarized by Theorems 2.2.1, 2.2.2, and 2.3.1 in this chapter. Note that for non-degenerate P-statistics ($r = 1$)

$$U_{g_1}^{(m,n)} = \binom{m}{1}^{-1}\binom{n}{1}^{-1}Per_{g_1}\mathbb{X}\,(1|1) = \frac{1}{mn}\sum_{ij}g_1(x_{ij})$$

and thus by Theorem 2.2.1

$$\frac{Per_h \, \mathbb{X}^{(m,n)}}{\binom{n}{m} m!} - Eh = \frac{1}{n} \sum_{ij} g_1(x_{ij}) + R_{m,n}$$

where the two summands are orthogonal by Theorem 2.2.2. To complete the proof it is enough then to show $R_{m,n} = o_p(m/\sqrt{n})$ which will follow if we can argue that

$$\frac{n}{m^2} Var \, R_{m,n} \to 0$$

as $n \to \infty$ and $m^2/n \to 0$. Under the assumptions of the example we have

$$Var \, U_{g_c}^{(m,n)} = \binom{n}{c}^{-1} \binom{m}{c}^{-1} \sum_{d=0}^{c} \binom{m-d}{c-d} \frac{D(c,d)}{d!} \leq \binom{n}{c}^{-1} \exp(M_2) \, M_1^c$$

since $\binom{m-d}{c-d} \leq \binom{m}{c}$ for $1 \leq d \leq c \leq m$. Therefore by Theorem 2.3.1

$$\frac{n}{m^2} Var \, R_{m,n} = \frac{n}{m^2} \sum_{c=2}^{m} \binom{m}{c}^2 Var \, U_{g_c}^{(m,n)} \leq \frac{n}{m^2} \exp(M_2) \sum_{c=2}^{m} \binom{m}{c}^2 \binom{n}{c}^{-1} M_1^c$$

$$\leq \frac{n}{m^2} \exp(M_2) \sum_{c=2}^{m} \left(M_1 \frac{m^2}{n} \right)^c \frac{1}{c!},$$

where the last inequality follows by

$$c! \frac{\binom{m}{c}^2}{\binom{n}{c}} \leq \left(\frac{m^2}{n} \right)^c$$

Consequently,

$$\frac{n}{m^2} Var \, R_{m,n} \leq M_1^2 \exp(M_2) \frac{m^2}{n} \sum_{c=0}^{m} \left(\frac{M_1 m^2}{n} \right)^c \frac{1}{c!} \to 0,$$

since the sum above is bounded by a constant and $m^2/n \to 0$.

In particular, continuing the case of random permanent (see Examples 2.2.1 and 2.3.1), we take $h(y_1, \ldots, y_m) = \prod_{i=1}^{m} (y_i/\mu)$ such that $Eh = \mu \neq 0$. Then $[g_1(X_{i,j})] = \mu^{-1}[X_{i,j} - \mu]$. Thus the above formula provides the approximation for the classical permanent function since then we may take (see the formula for $D(c,d)$ in Example 2.3.1) $M_1 = \rho \gamma^2$ and $M_2 = (1 - \rho)/\rho$.

2.4 Bibliographic Details

The standard general references on the martingale theory are Chow and Teicher (1978); Billingsley (1999) and for continuous time processes e.g.,

Ethier and Kurtz (1986). The general martingale decomposition for a U-statistic, which became latter known as the H-decomposition, was presented for the first time in Hoeffding (1961). The notion of a P-statistic and its general decomposition comes from the papers of Rempała and Wesołowski (2003) and Rempała (2001). The idea is related to a general canonical decomposition of a symmetric functional as described in Dynkin and Mandelbaum (1983) or in Lee (1990, chapter 4). Most of the material of this chapter discussed in Sections 2 and 3 comes from the work of Rempała (2001), which, though appeared earlier, extended to P-statistics the ideas developed for random permanents in Rempała and Wesołowski (2002b) and Rempała and Wesołowski (2002c).

3

Asymptotics for Random Permanents

3.1 Introduction

As already pointed out, the major difficulty of working with permanents and, in general, with P-statistics, lies in the fact that they are very difficult to compute when the dimensions of the matrix are large. In such case a natural question arises whether one could approximate the complicated exact distribution of a P-statistic by means of a simpler one. The issue leads to the problem of limiting behavior of P-statistics and particularly to the problem of weak convergence. In order to prepare the discussion of the problem for general P-statistics later on (Chapter 5), we shall first investigate in the next two chapters a somewhat simpler (but very insightful) case of a random permanent function.

In case of a random permanent the first results on weak convergence were obtained for one dimensional projection matrices in Székely (1982), van Es and Helmers (1988), Borovskikh and Korolyuk (1994). For the record we state them below as Theorem 3.1.1. In the sequel we use the symbol $\xrightarrow{d}$ to denote the convergence in distribution of random variables in $\mathbf{R}^k$ (or weak convergence of their distributions). A more general discussion of the notion of weak convergence in metric spaces is given in the next chapter. Before stating the first result let us recall that for a random variable X with mean $\mu \neq 0$ and variance $\sigma^2 < \infty$ its *coefficient of variation* is defined as $\gamma = \sigma/|\mu|$.

Theorem 3.1.1. *Let (X_l) be a sequence of iid positive random variables with a coefficient of variation $\gamma < \infty$. Let $\underline{X}$ be the corresponding $k \times l$ projection matrix $(k \leq l)$ and let $\mathcal{N}$ denote a standard normal random variable.*

(i) If $k^2/l \to 0$ then

$$\frac{\sqrt{l}}{k} \left(\frac{Per\, \underline{X}}{E\, Per\, \underline{X}} - 1 \right) \xrightarrow{d} \gamma \mathcal{N}$$

(ii) If $k^2/l \to \lambda > 0$ then

$$\frac{Per\,\underline{X}}{E\,Per\,\underline{X}} \xrightarrow{d} \exp\left(\gamma\sqrt{\lambda}\mathcal{N} - \frac{\lambda\gamma^2}{2}\right)$$

(iii) If $k^2/l \to \infty$ then no limit in distribution exists. That is, there exist no sequences of real numbers (a_l) and (b_l) such that $a_l(Per\,\underline{X} - b_l)$ converges in law to a non-degenerate random variable.

Proof. We only argue part (iii) as parts (i)-(ii) are special cases of Theorem 3.4.3 which shall be discussed later on. Let us first state the following auxiliary result regarding asymptotics of symmetric elementary polynomials (cf. Example 1.4.1 part 2).

Lemma 3.1.1. *Under the assumptions of Theorem 3.1.1 with $\sigma^2 = 1$ let $k/l \to 0$ as $l, k \to \infty$, then*

$$\sqrt{l}\left[\left(\binom{l}{k}^{-1}S_l(k)\right)^{1/k} - s_l\right] \xrightarrow{d} \mathcal{N}$$

where s_l is a sequence of positive numbers such that $s_l \to \mu$.

The above lemma is stated without proof which may be found in the original paper of Székely (1982). For the proof of the main result define now

$$V_l = \sqrt{l}\left[\left(\binom{l}{k}^{-1}S_l(k)\right)^{1/k} - s_l\right]$$

and note that in view of Lemma 3.1.1 the random variable V_l is asymptotically standard normal. Since $s_l + V_l/\sqrt{l} \geq 0$, Taylor's expansion (cf. also the proof of Theorem 7.2.1 in Chapter 7), yields

$$\binom{l}{k}^{-1}S_l(k) = \exp\left\{k\log\left(s_l + \frac{V_l}{\sqrt{l}}\right)\right\} = s_l^k \exp\left(\frac{k\,W_l}{\sqrt{l}}\right) \qquad (3.1)$$

where W_l is asymptotically normal with zero mean and variance μ^{-2}. Now, consider

$$Z_l = \exp\left(\frac{k\,W_l}{\sqrt{l}}\right)$$

and suppose that there exist sequences (a_l) and (b_l) with $a_l > 0$ such that $a_l(Z_l - b_l)$ converges in distribution to some non-degenerate random variable with a distribution function F. Then choose three continuity points u_1, u_2, u_3

of F such that $0 \leq F(u_1) < F(u_2) < F(u_3) \leq 1$. By the definition of Z_l we have for $i = 1, 2, 3$

$$\lim_l P(\{a_l(Z_l - b_l) \leq u_i\}) = \lim_l P\left(W_l \leq \sqrt{l}k^{-1}\log\left(\frac{u_i + a_l b_l}{a_l}\right)\right) = F(u_i).$$

Due to the uniform convergence of distribution functions in the central limit theorem this implies in turn that

$$\lim_l \sqrt{l}\, k^{-1}\left(\log\left(\frac{u_2 + a_l b_l}{a_l}\right) - \log\left(\frac{u_1 + a_l b_l}{a_l}\right)\right)$$
$$= \frac{1}{\mu}\left\{\Phi^{-1}(F(u_2))) - \Phi^{-1}(F(u_1))\right\} > 0$$

and thus

$$\lim_l \sqrt{l}\, k^{-1}\log\left(\frac{u_2 + a_l b_l}{u_1 + a_l b_l}\right) > 0.$$

Since $\sqrt{l}\, k^{-1} \to 0$ hence $a_l b_l \to -u_1$. But the same argument applied now to u_2 and u_3 gives also $a_l b_l \to -u_2$ and thus leads to a contradiction.   □

The above result indicates that the limiting behavior of the random permanent function depends heavily on the relationships between the dimensions of the matrix. We shall examine this issue in detail further on. Before proceeding, we shall formulate and prove several background results.

3.2 Preliminaries

3.2.1 Limit Theorems for Exchangeable Random Variables

We start the discussion with a quick review of some results on the sequences of exchangeable random variables. Recall that according to Definition 1.6.2 a sequence of random variables is exchangeable if the joint distribution of its every finite subsequence is invariant under permutations. It follows that a sequence of exchangeable random variables is always *stationary*.

Example 3.2.1 (Correlation). Consider the setting of Example 1.6.1 assuming additionally that $X_1 \in L^2(\Omega, \mathcal{F}, P)$. Note that since the sequence (X_n) is exchangeable it must have a common correlation coefficient ρ. Moreover,

$$\rho = Cov\,(X_1, X_2) = \frac{E(X_1 X_2) - EX_1 EX_2}{(Var\, X_1 \, Var\, X_2)^{1/2}}$$
$$= \frac{E(Y_1 Y_2 + Y Y_1 + Y Y_2 + Y^2) - (EY_1 + EY)(EY_2 + EY)}{Var\, X_1}$$
$$= \frac{EY^2 - (EY)^2}{Var\, Y_1 + Var\, Y} = \frac{Var\, Y}{Var\, Y_1 + Var\, Y}.$$

The above example motivates the consideration of bounds on the common correlation coefficient. Assume $Y_1 \ldots, Y_n$ are exchangeable with some common variance $\sigma^2 > 0$ and common correlation coefficient ρ. Then

$$0 \leq Var\left(\sum_{i=1}^{n} Y_i\right) = \sum_{i=1}^{n} Var\left(Y_i\right) + \sum_{i\neq j} Cov\left(Y_i, Y_j\right)$$
$$= \sigma^2 n(1 + (n-1)\rho)$$

which yields the bounds on ρ

$$\frac{-1}{n-1} \leq \rho \leq 1. \tag{3.2}$$

Thus it follows that for an infinite sequence of exchangeable random variables ρ has to be non-negative.

The basic result for exchangeable sequences is the celebrated de Finetti theorem. We state it without proof which may be found for instance in Chow and Teicher (1978, p. 226) or Taylor et al. (1985, chapter 2 p. 17)

Theorem 3.2.1 (de Finetti Theorem). *For an infinite sequence of exchangeable random variables* $(Y_n)_{n\geq1}$ *there exists a σ-algebra $\mathcal{G}$ such that*

$$P(Y_1 \leq y_1 \ldots, Y_m \leq y_m) = \int \prod_{i=1}^{m} P(Y_i \leq y_i | \mathcal{G}) \, dP$$

for all $y_1, \ldots, y_m \in \mathbf{R}$ *and* $m \geq 1$. *In the above we may take for $\mathcal{G}$ a tail σ-algebra of the sequence* (Y_n). $\square$

To illustrate the result let us consider a simple example.

Example 3.2.2 (Binary exchangeable sequence). Let $(Y_n)_{n\geq1}$ be an exchangeable sequence of variables taking values in $\{0, 1\}$. By the de Finetti theorem above there exists a distribution function F concentrated on $[0,1]$ such that

$$P(Y_1 = 1, \ldots, Y_k = 1, Y_{k+1} = 0, \ldots, Y_n = 0) = \int_0^1 \theta^k (1 - \theta)^{n-k} dF(\theta).$$

Looking at the above relationship from the Bayesian statistics point of view, the distribution F may be regarded as a Bayesian prior on the random parameter Θ. In fact given $\Theta = \theta$ the variables $Y_1, Y_2, \ldots$, are independent Bernoulli variates with parameter θ and thus the above relationship clearly holds.

One of the immediate consequences of the de Finetti theorem is a version of the central limit theorem for sums of exchangeable random variables. In order to state the result we need to introduce an additional definition.

Definition 3.2.1 (Mixture of normals). *Let Φ be a cumulative distribution function of a standard normal random variable. A random variable Z with distribution function F is a mixture of (centered) normals if F satisfies*

$$F(x) = \int_0^\infty \Phi(x/\sigma)dG(\sigma), \quad x \in \mathbf{R},$$

where G is a distribution function concentrated on $[0,\infty)$. When $\sigma = 0$ we put $\Phi(x/\sigma) = I(x \geq 0)$.

To clarify the above definition note that if Z is a random variable with distribution F as above, then Z is equal in distribution to a product of two independent random variables η and $\mathcal{N}$ where $\mathcal{N}$ is a standard normal variate and η has the distribution function G. Indeed, let $i = \sqrt{-1}$ and ϕ be the standard normal density function and consider the characteristic function of Z for $t \in \mathbf{R}$

$$E \exp(itZ) = \int \exp(itx)dF(x) = \int \exp(itx)d\left(\int \Phi(x/\sigma)dG(\sigma)\right)$$

$$= \int \exp(itx)\int \sigma^{-1}\phi(x/\sigma)dG(\sigma)dx$$

$$= \int\int \exp(it\sigma x)\phi(x)dx\,dG(\sigma) = E\exp(it\eta\mathcal{N}).$$

Below we state the versions of a central limit theorem and of a law of large numbers for the weighted sums of exchangeable random variables. For the proof of the central limit theorem result, see Taylor et al. (1985, chapter 2).

Theorem 3.2.2 (CLT for exchangeable sequences). *Let $(Y_n)_{n\geq 1}$ be an infinite sequence of exchangeable random variables such that $EY_1 = 0$, $0 < EX_1^2 = \sigma^2 < \infty$ and $Cov\,(Y_1, Y_2) = 0$. Let $[a_{nk} : 1 \leq k \leq k_n, n \geq 1]$ be a triangular array of real constants. Assume that the following two conditions hold:*

(i) $\max_{1\leq k\leq k_n} |a_{nk}| \to 0$,
(ii) $\sum_{k=1}^{k_n} a_{nk}^2 \to 1$,   as $n \to \infty$.

Then

$$\sum_{k=1}^{k_n} a_{nk}Y_k \xrightarrow{d} Z,$$

as $n \to \infty$ where the random variable Z is a mixture of normals.   □

Theorem 3.2.3 (SLLN for exchangeable sequences). *Let $(Y_n)_{n\geq 1}$ be an infinite sequence of exchangeable random variables such that $E|Y_1| < \infty$. Let $\mathcal{G}$ be the tail σ-field of the sequence $(Y_n)_{n\geq 1}$. Then*

$$\frac{1}{n}\sum_{k=1}^{n} Y_k \to E(Y_1|\mathcal{G}) \quad a.s.$$

□

For the proof see, e.g., Chow and Teicher (1978, chapter 7).

3.2.2 Law of Large Numbers for Triangular Arrays

In addition to basic limiting results for sequences of exchangeable random variables in the sequel we shall also make use of some law of large numbers type results for triangular arrays. Let us denote by $\{Y_{lk}\}$ for $l \geq 1$, $k = 1, \ldots, l$, an arbitrary triangular array of real random variables and consider the following conditions

(B1) $\sup_l \frac{1}{l} \sum_{k=1}^{l} E|Y_{lk}| = \beta < \infty$

(B2) $\frac{1}{l} \sum_{k=1}^{l} E|Y_{lk}|I\{|Y_{lk}| > c_l\} \to 0$ as $l \to \infty$, where the sequence (c_l) satisfies $c_l \to \infty$ and $c_l/l \to 0$.

Theorem 3.2.4 (WLLN for triangular arrays). *Let $\{Y_{lk}\}$ for $l \geq 1$, $k = 1, \ldots, l$, be a triangular array of row-wise independent, integrable random variables. If the conditions (B1)-(B2) are satisfied then as $l \to \infty$*

(i) $\frac{1}{l} \sum_{k=1}^{l} (Y_{lk} - E Y_{lk}) \xrightarrow{P} 0$

(ii) $\frac{1}{l^\alpha} \sum_{k=1}^{l} |Y_{lk}|^\alpha \xrightarrow{P} 0$ *for any $\alpha > 1$.*

Proof. For the proof of (i), let us consider the expression

$$\frac{1}{l} \sum_{k=1}^{l} (Y_{lk} - E Y_{lk}) = \frac{1}{l} \sum_{k=1}^{l} (Y_{lk}I\{|Y_{lk}| \leq c_l\} - E Y_{lk}I\{|Y_{lk}| \leq c_l\}) +$$

$$\frac{1}{l} \sum_{k=1}^{l} (Y_{lk}I\{|Y_{lk}| > c_l\} - E Y_{lk}I\{|Y_{lk}| > c_l\}) = (I) + (II) \qquad (3.3)$$

But in view of (B1) and (B2) the expressions (I) and (II) both converge to zero in probability. Indeed, apropos (I), for any $\varepsilon > 0$

$$P\left(\left|\frac{1}{l} \sum_{k=1}^{l} (Y_{lk}I\{|Y_{lk}| \leq c_l\} - E Y_{lk}I\{|Y_{lk}| \leq c_l\})\right| > \varepsilon\right)$$

$$\leq \frac{1}{\varepsilon^2 l^2} \sum_{k=1}^{l} E Y_{lk}^2 I\{|Y_{lk}| \leq c_l\} \leq \frac{c_l}{\varepsilon^2 l} \beta \to 0,$$

as $l \to \infty$ in view of (B1) and $c_l/l \to 0$.

Similarly, apropos (II), for any $\varepsilon > 0$ we have

$$P\left(\left|\frac{1}{l} \sum_{k=1}^{l} (Y_{lk}I\{|Y_{lk}| > c_l\} - E Y_{lk}I\{|Y_{lk}| > c_l\})\right| > \varepsilon\right)$$

$$\leq \frac{2}{l\varepsilon} \sum_{k=1}^{l} E|Y_{lk}|I\{|Y_{lk}| > c_l\} \to 0,$$

as $l \to \infty$, in view of the assumption (B2).

For the proof of (ii), let us consider a decomposition similar to (3.3)

$$\frac{1}{l^\alpha} \sum_{k=1}^{l} |Y_{lk}|^\alpha$$

$$= \frac{1}{l^\alpha} \sum_{k=1}^{l} |Y_{lk}|^\alpha I\{|Y_{lk}| \leq c_l\} + \frac{1}{l^\alpha} \sum_{k=1}^{l} |Y_{lk}|^\alpha I\{|Y_{lk}| > c_l\} = (III) + (IV).$$

Let $\varepsilon > 0$ be arbitrary. The expression (III) converges to zero in probability in view of the inequality

$$P\left(\left| \frac{1}{l^\alpha} \sum_{k=1}^{l} |Y_{lk}|^\alpha I\{|Y_{lk}| \leq c_l\} \right| > \varepsilon \right) \leq \left(\frac{c_l}{l} \right)^{\alpha-1} \frac{\beta}{\varepsilon}$$

and the expression (IV) converges to zero in probability in view of

$$P\left(\left| \frac{1}{l^\alpha} \sum_{k=1}^{l} |Y_{lk}|^\alpha I\{|Y_{lk}| > c_l\} \right| > \varepsilon \right) \leq P\left(\sum_{k=1}^{l} \frac{|Y_{lk}|}{l} I\{|Y_{lk}| > c_l\} > \varepsilon^{1/\alpha} \right)$$

$$\leq \frac{1}{\varepsilon^{1/\alpha} l} \sum_{k=1}^{l} E\,|Y_{lk}| I\{|Y_{lk}| > c_l\}$$

and the condition (B2). $\qquad\qquad\square$

3.2.3 More on Elementary Symmetric Polynomials

In Chapter 1 we have defined an elementary symmetric polynomial $S_l(c)$ (cf. Example 1.4.1). Note that it can be written as $S_l(c) = \pi_l^c[h]$, where h is defined by $h(y_1, \ldots, y_c) = y_1 \ldots\ldots y_c$. For the sake of notational convenience we also define $S_l(0) \equiv 1$ and take the sum in the definition of $S_l(c)$ to be zero whenever $c > l$. Elementary symmetric polynomials have several interesting properties. For instance, for any l the polynomials $S_l(c)$ satisfy the following recursive relationship.

Lemma 3.2.1. *For any* $1 \leq c \leq l$

$$c\,S_l(c) = \sum_{d=0}^{c-1} (-1)^d S_l(c-d-1) \left(\sum_{k=1}^{l} y_k^{d+1} \right). \tag{3.4}$$

Proof. Let us introduce an auxiliary polynomial $Q(c, d)$ defined as

$$Q(c,d)$$

$$= \sum_{1 \leq k_1 < \ldots < k_c \leq l} y_{k_1}^d \cdots y_{k_c} + \sum_{1 \leq k_1 < \ldots < k_c \leq l} y_{k_1} y_{k_2}^d \cdots y_{k_c} + \ldots + \sum_{1 \leq k_1 < \ldots < k_c \leq l} y_{k_1} \cdots y_{k_c}^d$$

$$= \sum_{k=1}^{l} y_k^d \left(\sum_{k_i \neq k} y_{k_1} \cdots y_{k_{c-1}} \right),$$

where in the last expression the inner summation is taken over all possible choices of $c - 1$ distinct indices $k_1, \ldots, k_{c-1}$ out of the set of indices $\{1, 2, \ldots, l\} \backslash \{k\}$. From the definition of $Q(c, d)$ it follows that for any positive integers $1 \leq c \leq l$ and d

$$Q(c, 1) = c\, S_l(c) \quad \text{and} \quad Q(1, d) = \sum_{k=1}^{l} y_k^d, \tag{3.5}$$

as well as that

$$Q(c, d) + Q(c - 1, d + 1) = S_l(c - 1) \left(\sum_{k=1}^{l} y_k^d \right). \tag{3.6}$$

Now, solving the recursion (3.6) for $Q(c, 1)$ with the help of the identities (3.5) we arrive at (3.4). □

The following concept of (monic) Hermité polynomials often arises naturally in connection with the elementary symmetric polynomials.

Definition 3.2.2 (Hermité polynomials). *The (monic) Hermité polynomials (H_n) are defined by the formula*

$$\forall_{s,t} \sum_{n=0}^{\infty} \frac{t^n}{n!} H_n(s) = \exp\left(st - \frac{t^2}{2}\right).$$

From the definition it follows that

$$H_n(s) = \frac{d^n}{dt^n} \bigg|_{t=0} \exp\left(st - \frac{t^2}{2}\right)$$

which gives

$$\begin{aligned} H_1(s) &= s; \\ H_2(s) &= s^2 - 1; \\ H_3(s) &= s^3 - 3\,s; \\ H_4(s) &= s^4 - 6s^2 + 3; \end{aligned}$$

and so on. In fact, it can be also shown that

$$H_{n+1}(s) = s\, H_n(s) - n\, H_{n-1}(s). \tag{3.7}$$

Other relations between Hermité polynomial include for instance

$$H_n(s) = (-1)^n \exp\left(\frac{s^2}{2}\right) \frac{d^n}{ds^n} \exp\left(-\frac{s^2}{2}\right)$$

and
$$\frac{d}{ds} H_n(s) = n\, H_{n-1}(s).$$

However, the most important property of Hermité polynomial is that they are orthogonal in the space $L^2(\mathbf{R}, \mathcal{B}, P_{\mathcal{N}})$ of all square integrable functions of standard normal variable $\mathcal{N}$. More precisely,

$$E[H_m(\mathcal{N})H_n(\mathcal{N})] = \frac{1}{\sqrt{2\pi}} \int H_m(s)\, H_n(s)\, e^{-s^2/2} \, ds = \begin{cases} 0 & \text{if } m \neq n, \\ n! & \text{otherwise.} \end{cases}$$

3.3 Limit Theorem for Elementary Symmetric Polynomials

Of course the definition of elementary symmetric polynomials works for arguments coming from an array of random variables as well. We state it for the reference below. For an arbitrary triangular array $\{Y_{lk}\}$ $(1 \leq k \leq l)$, of real random variables a corresponding elementary symmetric polynomial statistic of order $c \geq 1$ is defined as

$$S_l(c) = \sum_{1 \leq i_1 < \ldots < i_c \leq l} Y_{l,i_1} \cdots Y_{l,i_c}. \tag{3.8}$$

We shall start with a result on the weak convergence of elementary symmetric polynomials based on triangular arrays of real random variables.

Theorem 3.3.1. *Let $\{Y_{lk}\}$ be a triangular array of square integrable, row-wise independent, real random variables such that $E\,Y_{lk} = 0$ and*

$$\lim_{l \to \infty} \frac{1}{l}\, Var\left(\sum_{k=1}^{l} Y_{lk}\right) = \sigma^2 > 0,$$

satisfying additionally the Lindeberg condition

(LC) $\forall_{\varepsilon > 0}\ \frac{1}{l} \sum_{k=1}^{l} E\,Y_{lk}^2\, I\{|Y_{lk}| > \sigma\sqrt{l}\,\varepsilon\} \to 0 \quad$ as $\quad l \to \infty.$

Then, for any integer $c \geq 1$,

$$\left[\frac{S_l(1)}{l^{1/2}}, \frac{S_l(2)}{l^{2/2}}, \ldots, \frac{S_l(c)}{l^{c/2}}\right]^T \xrightarrow{d} \left[\frac{\sigma\, H_1(\mathcal{N})}{1!}, \frac{\sigma^2\, H_2(\mathcal{N})}{2!}, \ldots, \frac{\sigma^c\, H_c(\mathcal{N})}{c!}\right]^T \tag{3.9}$$

where H_c is the Hermité polynomial of order c and $\mathcal{N}$ is a standard normal random variable.

Proof. Let us assume (without loosing generality) that $\sigma^2 = 1$. We shall prove the above result by induction with respect to c. For $c = 1$ the result is just the classical CLT (see Corollary 4.3.1) since we have assumed the required condition *(LC)* and since $H_1(\mathcal{N}) = \mathcal{N}$. Hence

$$\frac{S_l(1)}{l^{1/2}} \xrightarrow{d} H_1(\mathcal{N}) \tag{3.10}$$

Let us assume thus that (3.9) holds true for all integers $i = 1, \ldots, c-1$ and define a $(c-1)$-dimensional vector

$$\mathbb{Y}_c^{(l)} = \left[\frac{\sum Y_{lk}^2}{l^{2/2}}, \frac{\sum Y_{lk}^3}{l^{3/2}}, \ldots, \frac{\sum Y_{lk}^c}{l^{c/2}} \right]^T.$$

First, let us note that by taking $X_{lk} = Y_{lk}^2$ we obtain the row-wise independent random array satisfying the conditions (B1)-(B2) of the Theorem 3.2.4 with $\beta = 1$ and $c_l = \sqrt{l}$. Thus,

$$\mathbb{Y}_c^{(l)} \xrightarrow{P} [1, 0, \ldots, 0]^T \quad \text{as} \quad l \to \infty. \tag{3.11}$$

But (3.10) along with (3.11) imply that (cf. e.g., Billingsley 1999, Chap. 1) the vector $\left[S_l(1)/\sqrt{l}, \mathbb{Y}_c^{(l)} \right]^T$ converges weakly to the corresponding limit. On the other hand, by (3.4) we have that for $i = 1, \ldots, c-1$

$$l^{-i/2} S_l(i) = G_i \left(\frac{S_l(1)}{l^{1/2}}, \mathbb{Y}_c^{(l)} \right)$$

where the continuous function G_i (known as the Waring function) depends only upon i but not upon l. This and the induction hypothesis imply

$$\begin{bmatrix} \frac{S_l(1)}{l^{1/2}} \\ \vdots \\ \frac{S_l(c-1)}{l^{(c-1)/2}} \\ \frac{S_l(c)}{l^{c/2}} \end{bmatrix} = \begin{bmatrix} G_1 \left(\frac{S(1)}{l^{1/2}}, \mathbb{Y}_c^{(l)} \right) \\ \vdots \\ G_{c-1} \left(\frac{S(1)}{l^{1/2}}, \mathbb{Y}_c^{(l)} \right) \\ G_c \left(\frac{S(1)}{l^{1/2}}, \mathbb{Y}_c^{(l)} \right) \end{bmatrix} \xrightarrow{d} \begin{bmatrix} H_1(\mathcal{N})/1! \\ \vdots \\ H_{c-1}(\mathcal{N})/(c-1)! \\ B_c \end{bmatrix} \tag{3.12}$$

for some random variable B_c. Hence, in order to arrive at (3.9), we need only to identify B_c. To this end, using formula (3.4) we write

$$\frac{S_l(c)}{l^{c/2}} = \frac{1}{c} \left[\frac{S_l(c-1)}{l^{(c-1)/2}} \frac{S_l(1)}{l^{1/2}} - \frac{S_l(c-2)}{l^{(c-2)/2}} \frac{\sum_{k=1}^l Y_{lk}^2}{l} \right] + A_c.$$

By (3.11) and (3.12), the first part of the above right hand side converges to

$$\frac{H_{c-1}(\mathcal{N}) \mathcal{N} - (c-1) H_{c-2}(\mathcal{N})}{c!} = \frac{H_c(\mathcal{N})}{c!},$$

in view of the recurrence relation for Hermité polynomials given by (3.7). For the reminder term A_c, we have

$$A_c = \frac{1}{c} \sum_{d=2}^{c-1} (-1)^d \frac{S_l(c-1-d)}{l^{(c-1-d)/2}} \left(\sum_{k=1}^{l} \frac{Y_{lk}^{2(d+1)/2}}{l^{(d+1)/2}} \right) \xrightarrow{P} 0$$

since the expressions in paranthesis converges in probability to zero (by Theorem 3.2.4) and the random coefficients standing in front of each of these expressions converge in distribution due to the induction assumption. The above implies that $B_c = H_c(\mathcal{N})/c!$, which entails (3.9). □

The result above implies in particular that

$$\frac{S(c)}{l^{c/2}} \xrightarrow{d} \frac{H_c(\mathcal{N})}{c!},$$

which is a result derived in Teicher (1988) and a special case of the quite general limit theorem for U-statistics based on arrays of random variables of Rubin and Vitale (1980). We shall consider a more general version of that result later on.

3.4 Limit Theorems for Random Permanents

Having established the results on elementary symmetric polynomials in the previous sections, let us turn to the problem of weak convergence of random permanents. For convenience we briefly recap the basic notions and notation introduced in Chapter 2. As before we (for now) assume that $\mathbb{X}^{(m,n)} = [X_{ij}]$ is an $m \times n$ ($m \le n$) real random matrix of square integrable components coming from $\mathbb{X}^{(\infty)}$ which satisfies the exchangeability conditions (A1-A2). Clearly, under these assumptions all entries of the matrix $\mathbb{X}$ are identically distributed although not necessarily independent. For $i, k = 1, \ldots, m$ and $j = 1, \ldots, n$ we denote $\mu = E\, X_{ij}$, $\sigma^2 = Var\, X_{ij}$ and $\rho = Corr(X_{kj}, X_{ij}) > 0$ (see (3.2)). In what follows, we shall always assume that $\mu \ne 0$ and we shall also denote by $\gamma = \sigma/|\mu|$ the coefficient of variation.

The major tool of the investigation shall be the orthogonal decomposition described in Example 2.2.1. Recall that

$$\frac{Per\, \mathbb{X}^{(m,n)}}{\binom{n}{m} m! \mu^m} = 1 + \sum_{c=1}^{m} \binom{m}{c} U_c^{(m,n)}, \qquad (3.13)$$

where

$$U_c^{(m,n)} = \binom{n}{c}^{-1} \binom{m}{c}^{-1} c!^{-1} Per\tilde{\mathbb{X}}(c|c),$$

for $\tilde{\mathbb{X}}^{\infty} = [X_{ij}/\mu - 1]$. Moreover, from Theorem 2.3.1 and Example 2.3.1 we know that under the assumptions on the entries of $\mathbb{X}^{(\infty)}$ the random variables $U_c^{(m,n)}$ above for $c = 1, 2 \ldots, m$, are orthogonal, i.e.

$$Cov\left(U_{c_1}^{(m,n)}, U_{c_2}^{(m,n)}\right) = 0 \qquad \text{for } c_1 \neq c_2 \tag{3.14}$$

with the variance

$$Var\, U_c^{(m,n)} = \binom{n}{c}^{-1}\binom{m}{c}^{-1}\gamma^{2c}\sum_{r=0}^{c}\binom{m-r}{c-r}\frac{\rho^{c-r}(1-\rho)^r}{r!}. \tag{3.15}$$

Let us note that when $\rho = 1$ then $\mathbb{X}^{(\infty)} = \underline{X}$ and $\frac{Per\,\mathbb{X}^{(m,n)}}{\binom{n}{m}m!\mu^m}$ reduces to a P-statistic with the kernel as in Example 2.2.1, that is, a normalized elementary symmetric polynomial in variables X_i/μ.

For any c and suitably large m and n denote

$$W_c(n) = \binom{n}{c}\binom{m}{c}c!\,U_c^{(m,n)} = Per\,\tilde{\mathbb{X}}\,(c|c)$$

$$= \sum_{1 \leq i_1 < \ldots < i_c \leq m}\;\sum_{1 \leq j_1 < \ldots < j_c \leq n}Per\,\tilde{\mathbb{X}}\,(i_1, \ldots, i_c | j_1, \ldots, j_c).$$

The following results describes the asymptotic behavior of $W_c(n)$ which is the key to investigating the asymptotic behavior of the decomposition (3.13).

Theorem 3.4.1. *Let c be an arbitrary positive integer. Assume that $m = m_n \to \infty$ as $n \to \infty$.*
If $\rho = 0$ then

$$\left[\frac{W_1(n)}{(\sqrt{nm})}, \ldots, \frac{W_c(n)}{(\sqrt{nm})^c}\right]^T \xrightarrow{d} \left[\frac{\gamma}{1!}H_1(\mathcal{N}), \ldots, \frac{\gamma^c}{c!}H_c(\mathcal{N})\right]^T,$$

If $\rho > 0$ then

$$\left[\frac{W_1(n)}{\sqrt{nm}}, \ldots, \frac{W_c(n)}{(\sqrt{nm})^c}\right]^T \xrightarrow{d} \left[\frac{\gamma\rho^{1/2}}{1!}H_1(\mathcal{N}), \ldots, \frac{\gamma^c\rho^{c/2}}{c!}H_c(\mathcal{N})\right]^T.$$

Proof. Consider first the case $\rho = 0$. Define for an arbitrary fixed positive integer c

$$V_c(n) = n^{-c/2}\sum_{1 \leq j_1 < \ldots < j_c \leq n}Y_{n,j_1}\ldots Y_{n,j_c},$$

where

$$Y_{n,j} = \frac{1}{\sqrt{m}}\sum_{i=1}^{m}\tilde{X}_{ij},$$

$j = 1, 2, \ldots, n$. Here the first subscript indicates the dependence on n through $m = m(n)$. Observe that $V_c(n)$ is just an elementary symmetric polynomial in independent identically distributed random variables which are column sums of the matrix $\tilde{\mathbb{X}}^{(m,n)}$ normalized by $\sqrt{m}$. Thus, $E(Y_{n,1}^2) = \gamma^2$ and we may use the result on the convergence of elementary symmetric polynomials for the row-wise independent identically distributed double arrays of square integrable random variables given in Theorem 3.3.1 to conclude that

$$[V_1(n), \ldots, V_c(n)]^T \xrightarrow{d} \left[\frac{\gamma}{1!}H_1(\mathcal{N}), \ldots, \frac{\gamma^c}{c!}H_c(\mathcal{N})\right]^T$$

as long as the following version of the Lindeberg condition holds:

$$E\left(Y_{n,1}^2 I(|Y_{n,1}| > \sqrt{n}\varepsilon)\right) \to 0,$$

for any $\varepsilon > 0$ as $n \to \infty$. To prove that the condition is satisfied observe first that

$$E\left(Y_{n,1}^2 I(|Y_{n,1}| > \sqrt{n}\varepsilon)\right) \le \sup_{k \ge 1} E\left(Y_{k,1}^2 I(Y_{k,1}^2 > n\varepsilon^2)\right).$$

Consequently, it suffices to show that the sequence of random variables $(Y_{k,1}^2)_{k \ge 1}$ is uniformly integrable, that is

$$\lim_{\alpha \to \infty} \sup_{k \ge 1} E\left(Y_{k,1}^2 I(Y_{k,1}^2 > \alpha)\right) = 0.$$

To this end, let us observe the following.

(i) By the de Finetti theorem (Theorem 3.2.1) and the central limit theorem for exchangeable sequences (Theorem 3.2.2) along with the fact that continuous mappings preserve weak convergence, it follows that $Y_{k,1}^2$ converges in distribution to $E(\tilde{X}_{1,1}^2|\mathcal{G})\mathcal{N}^2$, where $\mathcal{G}$ is the tail σ-field for the exchangeable sequence $(\tilde{X}_{i,1})_{i \ge 1}$, and $\mathcal{N}$ is a standard normal random variable independent of $\mathcal{G}$, and

(ii) $E(E(\tilde{X}_{1,1}^2|\mathcal{G})\mathcal{N}^2) = E(\tilde{X}_{1,1}^2) = \gamma^2$ which, on the other hand equals $E(Y_{k,1}^2)$ for any $k \ge 1$.

Finally, using the result of Lemma 3.4.1 below we conclude that the sequence $\{Y_{k,1}^2\}_{k \ge 1}$ is uniformly integrable since it converges in distribution and the corresponding sequence of expectations (all being equal) also converges to the suitable limit.

Observe that for any $k = 1, \ldots, c$

$$\frac{W_k(n)}{(\sqrt{nm})^k} = V_k(n) + \frac{R_k(n)}{(\sqrt{mn})^k},$$

where $R_k(n)$ is a sum of different products $\tilde{X}_{i_1,j_1} \ldots \tilde{X}_{i_k,j_k}$ such that $1 \le j_1 < \ldots j_k \le n$, $(i_1, \ldots, i_k) \in \{1, \ldots, m\}^k$ and at least one of $i_1, \ldots, i_k$ in the sequence $(i_1, \ldots, i_k)$ repeats.

Using the fact that $R_k(n)$ is a sum of orthogonal products (observe that the covariance of any two of such different products equals zero since the columns are independent and elements in each column have zero correlation) and that the variance of any of such single product equals γ^{2k} while the number of products in $R_k(n)$ equals $\binom{n}{k}(m^k - \binom{m}{k}k!)$ we obtain

$$Var\left(\frac{R_k(n)}{(\sqrt{nm})^k}\right) = \gamma^2 \frac{\binom{n}{k}(m^k - \binom{m}{k}k!)}{n^k m^k} \leq \frac{\gamma^2}{k!} \frac{m^k - \binom{m}{k}k!}{m^k} = O(1/m) \to 0,$$

as $n \to \infty$ (since the numerator is of order m^{k-1} while the denominator is of the order m^k).

Consequently, for $\rho > 0$ the assertion of the proposition follows.

Now, let us consider the case $\rho > 0$. Similarly as above, let us define

$$V_c(n) = n^{-c/2} \sum_{1 \leq j_1 < \ldots < j_c \leq n} Y_{n,j_1} \ldots Y_{n,j_c},$$

where

$$Y_{n,j} = \frac{1}{m} \sum_{i=1}^{m} \tilde{X}_{ij},$$

$j = 1, 2, \ldots, n$.

Similarly as in the first case, we will show that the sequence $(Y_{k,1}^2)_{k \geq 1}$ is uniformly integrable. To this end we observe first that by a version of law of large numbers for exchangeable sequences (Theorem 3.6.3) it follows that as $k \to \infty$

$$Y_{k,1}^2 \xrightarrow{d} E^2(\tilde{X}_{1,1}|\mathcal{G}),$$

where $\mathcal{G}$ is the tail σ-field. Further, by the de Finetti theorem (Theorem 3.2.1) we have

$$E(E^2(\tilde{X}_{1,1}|\mathcal{G})) = E(E(\tilde{X}_{1,1}|\mathcal{G})E(\tilde{X}_{2,1}|\mathcal{G})) = E(E(\tilde{X}_{1,2}\tilde{X}_{2,1}|\mathcal{F}))$$
$$= E(\tilde{X}_{1,1}\tilde{X}_{1,2}) = \rho\gamma^2.$$

Also,

$$E(Y_{k,1}^2) = \frac{1}{m^2}[m\gamma^2 + m(m-1)\rho\gamma^2] \to \rho\gamma^2, \quad k \to \infty$$

since $m = m(k) \to \infty$. Thus, the sequence $(Y_{k,1}^2)$ is uniformly integrable and the Lindeberg condition of Theorem 3.3.1 is satisfied.

This allows us to conclude that

$$[V_1(n), \ldots, V_c(n)]^T \xrightarrow{d} \left[\frac{(\sqrt{\rho}\gamma)^1}{1!}H_1(\mathcal{N}), \ldots, \frac{(\sqrt{\rho}\gamma)^c}{c!}H_c(\mathcal{N})\right]^T.$$

Similarly to the first case, for any $k = 1, \ldots, c$, we have

$$\frac{W_k(n)}{m^k(\sqrt{n})^k} = V_k(n) + \frac{R_k(n)}{m^k(\sqrt{n})^k}.$$

However, this time some of the elements of $R_k(n)$ are correlated - this is true for pairs of products originating from exactly the same columns; if at least one column in the pair of products is different then their correlation is zero. Consequently,

$$Var\left(\frac{R_k(n)}{m^k\sqrt{n}^k}\right) = \frac{1}{m^{2k}n^k}\sum_{1\le j_1<\cdots j_k\le n} Var\left(R_{j_1,\ldots,j_k}(n)\right)$$

$$= \frac{\binom{n}{k}}{m^{2k}n^k} Var\left(R_{1,\ldots,k}(n)\right),$$

where $R_{j_1,\ldots,j_k}(n)$ denotes the sum of respective products arising from the columns $j_1,\ldots,j_k$. Since

$$|Cov(\tilde{X}_{i_1,j_1}\cdots\tilde{X}_{i_k,j_k},\tilde{X}_{l_1,j_1}\cdots\tilde{X}_{l_k,j_k})|\le\gamma^{2k}$$

for any choices of rows $(i_1,\ldots,i_k)$ and $(l_1,\ldots,l_k)$, we conclude that

$$Var\left(R_{1,\ldots,k}(n)\right) < \left(m^k-\binom{m}{k}k!\right)^2\gamma^{2k}.$$

Hence, it follows that

$$Var\left(\frac{R_k(n)}{m^k(\sqrt{n})^k}\right) < \frac{\binom{n}{k}(m^k-\binom{m}{k}k!)^2\gamma^{2k}}{m^{2k}n^k} < \frac{\gamma^2}{k!}\frac{(m^k-\binom{m}{k}k!)^2}{m^{2k}}\to 0$$

as $n\to\infty$. Consequently,

$$\frac{R_k(n)}{n^{c/2}m^c}\xrightarrow{P} 0$$

and the assertion of the proposition for $\rho > 0$ follows. The proof of the result is thus complete as soon as we argue the assertion of Lemma 3.4.1.   □

Lemma 3.4.1. *Suppose that the sequence (Y_n) of positive integrable random variables satisfies $Y_n\xrightarrow{d} Y$ as $n\to\infty$ as well as $EY_n\to EY$ as $n\to\infty$ for some integrable random variable Y. Then the sequence Y_n is uniformly integrable.*

Proof. Consider

$$EY_n = \int_0^\alpha P(t < Y_n < \alpha)\,dt + \int_{Y_n\ge\alpha} Y_n dP$$

$$EY = \int_0^\alpha P(t < Y < \alpha)\,dt + \int_{Y\ge\alpha} Y\,dP.$$

By the hypothesis of the lemma for α such that $P(Y = \alpha) = 0$ we have

$$\int_{Y_n \geq \alpha} Y_n dP \to \int_{Y \geq \alpha} Y dP \quad as \ n \to \infty.$$

Thus for any $\varepsilon > 0$ with sufficiently large α and n_0 for all $n > n_0$ we must have

$$\int_{Y_n \geq \alpha} Y_n dP < \varepsilon.$$

Now, we may still increase α to ensure that $\max_{1 \leq n \leq n_0} \int_{Y_n \geq \alpha} Y_n dP < \varepsilon$ which implies that for sufficiently large α

$$\sup_n \int_{Y_n \geq \alpha} Y_n dP < \varepsilon$$

and the result follows.   □

The application of the martingale decomposition (3.13) along with Theorem 3.4.1 from the previous sections allow us to finally formulate main results for the asymptotic behavior of random permanents. We state them in Theorems 3.4.2 and 3.4.3 below, covering the cases $\rho = 0$ and $\rho > 0$, respectively. The first result incorporates the case of all-independent and identically distributed entries of $\mathbb{X}^{(\infty)}$. The second one treats, among others, the case of a one dimensional projection matrix ($\rho = 1$).

Theorem 3.4.2. *Assume that $\rho = 0$.*
   If $m/n \to \lambda > 0$ as $n \to \infty$ then

$$\frac{Per\left(\mathbb{X}^{(m,n)}\right)}{\binom{n}{m} m! \mu^m} \xrightarrow{d} \exp\left(\sqrt{\lambda}\gamma \mathcal{N} - \frac{\lambda\gamma^2}{2}\right).$$

If $m/n \to 0$ and $m = m_n \to \infty$ as $n \to \infty$ then

$$\sqrt{\frac{n}{m}}\left(\frac{Per\left(\mathbb{X}^{(m,n)}\right)}{\binom{n}{m} m! \mu^m} - 1\right) \xrightarrow{d} \gamma \mathcal{N}. \tag{3.16}$$

Proof. Consider first the case $\lambda > 0$. For any n and any N such that $N < m_n$ denote

$$S_{N,n} = 1 + \sum_{c=1}^{N} \binom{m}{c} U_c^{(m,n)} = 1 + \sum_{c=1}^{N} \frac{W_c(n)}{\binom{n}{c} c!} = 1 + \sum_{c=1}^{N} \frac{(\sqrt{nm})^c}{\binom{n}{c} c!} \frac{W_c(n)}{(\sqrt{nm})^c}.$$

Observe that by the first assertion of Proposition 3.4.1 we have that

$$S_{N,n} \xrightarrow{d} G_N = \sum_{c=0}^{N} \frac{(\lambda\gamma^2)^{c/2}}{c!} H_c(\mathcal{N}),$$

as $n \to \infty$, since for any $c = 1, 2, \ldots$

$$\frac{(\sqrt{nm})^c}{\binom{n}{c}c!} \to \sqrt{\lambda}.$$

Let us define also

$$T_{N,n} = \sum_{c=N+1}^{m_n} \binom{m}{c} U_c^{(m,n)}.$$

and observe that since $U_c^{(m,n)}$ are orthogonal then

$$Var\,(T_{N,n}) = \sum_{c=N+1}^{m_n} \frac{\binom{m}{c}}{\binom{n}{c}c!}\gamma^{2c} \le \sum_{c=N+1}^{\infty} \frac{\gamma^{2c}}{c!} = a_N,$$

and $a_N \to 0$ as $N \to \infty$. Consequently, for $Z_n = \frac{Per\,(\mathbb{X}^{(m,n)})}{\binom{n}{m}m!\mu^m}$ we have for any $\varepsilon > 0$

$$
\begin{aligned}
P(Z_n \le x) &= P(S_{N,n} + T_{N,n} \le x) \\
&\le P(S_{N,n} \le x + \varepsilon,\, |T_{N,n}| \le \varepsilon) + P(|T_{N,n}| > \varepsilon) \\
&\le P(S_{N,n} \le x + \varepsilon) + P(|T_{N,n}| > \varepsilon).
\end{aligned}
$$

On the other hand,

$$P(Z_n \le x) \ge P(S_{N,n} \le x - \varepsilon,\, |T_{N,n}| \le \varepsilon) \ge P(S_{N,n} \le x - \varepsilon) - P(|T_{N,n}| > \varepsilon).$$

Thus, for any $x \in \mathbf{R}$ and any $\varepsilon > 0$ we obtain the double inequality

$$
\begin{aligned}
P(S_{N,n} \le x - \varepsilon) &- P(|T_{N,n}| > \varepsilon) \le P(Z_n \le x) \le P(S_{N,n} \le x + \varepsilon) \\
&+ P(|T_{N,n}| > \varepsilon).
\end{aligned}
$$

Hence, by the Tchebyshev inequality it follows that

$$P(S_{N,n} \le x - \varepsilon) - a_N/\varepsilon^2 \le P(Z_n \le x) \le P(S_{N,n} \le x + \varepsilon) + a_N/\varepsilon^2.$$

Taking the limit as $n \to \infty$ we obtain

$$P(G_N \le x - \varepsilon) - a_N/\varepsilon^2 \le \lim_{n\to\infty} P(Z_n \le x) \le P(G_N \le x + \varepsilon) + a_N/\varepsilon^2.$$

But $a_N \to 0$ and G_N converges almost surely to $G_\infty = \exp(\sqrt{\lambda}\gamma\mathcal{N} - \lambda\gamma^2/2)$ as $N \to \infty$ since by the definition of Hermité polynomials (see, Definition 3.2.2)

$$\exp\left(\sqrt{\lambda}\gamma\mathcal{N} - \frac{\lambda\gamma^2}{2}\right) = \sum_{c=0}^{\infty} \frac{(\lambda\gamma^2)^{c/2}}{c!} H_c(\mathcal{N}).$$

Hence, for any $\varepsilon > 0$

$$P(G_\infty \leq x - \varepsilon) \leq \lim_{n \to \infty} P(Z_n \leq x) \leq P(G_\infty \leq x + \varepsilon).$$

Consequently, Z_n converges in distribution to G_∞, which completes the case $\lambda > 0$.

For the proof in the case $\lambda = 0$, let us write

$$\sqrt{\frac{n}{m}} \left(\frac{Per\,(\mathbb{X})}{\binom{n}{m}m!\mu^m} - 1 \right) = \sqrt{\frac{n}{m}} \binom{m}{1} U_1^{(m,n)} + R_{m,n} = \frac{W_1(n)}{\sqrt{nm}} + R_{m,n},$$

where

$$R_{m,n} = \sqrt{\frac{n}{m}} \sum_{c=2}^m \binom{m}{c} U_c^{(m,n)}.$$

Observe that $R_{m,n}$ converges in probability to zero, since by (3.15) it follows that

$$Var\,R_{m,n} = \frac{n}{m} \sum_{c=2}^m \frac{\binom{m}{c}\gamma^{2c}}{\binom{n}{c}c!} \leq \frac{m}{n} \exp(\gamma^2) \to 0.$$

Hence, the result follows by Theorem 3.4.1 for $m = m_n \to \infty$.

As an example of an application of this result consider

Example 3.4.1 (Asymptotics for the number of perfect matchings). Let $\mathbb{X}$ be a matrix of independent Bernoulli variables with probability of success p. Consider a number of perfect matchings $\mathcal{H}(G, \mathbb{X})$ in an undirected bipartite random graph $G = (V_1, V_2; E)$ as described in Example 1.3.1. Observe that in the notation of this chapter, for the Bernoulli random variables X_{ij} we have $\mu = p \in (0,1)$, $\gamma = \sqrt{(1-p)/p}$, and $\mathcal{H}(G, \mathbb{X}^{(m,n)}) = Per\,\mathbb{X}^{(m,n)}$. From Theorem 3.4.3 it follows that if $n, m \to \infty$ and $m/n \to \lambda > 0$ then

$$\frac{\mathcal{H}(G, \mathbb{X}^{(m,n)})}{\binom{n}{m}m!p^m} \xrightarrow{d} \exp\left(\sqrt{\lambda}\gamma\mathcal{N} - \frac{\lambda\gamma^2}{2} \right).$$

On the other hand, if $n - m \to \infty$ and $m/n \to 0$ then,

$$\sqrt{\frac{mp}{n(1-p)}} \left(\frac{\mathcal{H}(G, \mathbb{X}^{(m,n)})}{\binom{n}{m}m!p^m} - 1 \right) \xrightarrow{d} \mathcal{N}.$$

In the case $m = n$ the first of the above relations has been first noted, in a slightly different form, by Janson (1994).

Our second result on the convergence of random permanents treats the case $\rho > 0$ and, in particular, for $\rho = 1$ specializes to the parts (i)–(ii) of the Theorem 3.1.1.

Theorem 3.4.3. *Assume that $\rho > 0$.*
If $m^2/n \to \lambda > 0$ as $n \to \infty$ then

$$\frac{Per\left(\mathbb{X}^{(m,n)}\right)}{\binom{n}{m}m!\mu^m} \xrightarrow{d} \exp\left(\sqrt{\lambda\rho}\,\gamma\mathcal{N} - \frac{\lambda\rho\gamma^2}{2}\right). \tag{3.17}$$

If $m^2/n \to \lambda = 0$ and $m = m_n \to \infty$ as $n \to \infty$ then

$$\frac{\sqrt{n}}{m}\left(\frac{Per\left(\mathbb{X}^{(m,n)}\right)}{\binom{n}{m}m!\mu^m} - 1\right) \xrightarrow{d} \sqrt{\rho}\,\gamma\mathcal{N}. \tag{3.18}$$

Proof. As before, let us first consider the case $\lambda > 0$. Then for any N and any n such that $N < m_n$ denote

$$S_{N,n} = 1 + \sum_{c=1}^{N}\binom{m}{c}U_c^{(m,n)} = 1 + \sum_{c=1}^{N}\frac{(\sqrt{n}m)^c}{\binom{n}{c}c!}\frac{W_c(n)}{(\sqrt{n}m)^c}.$$

Observe that by the second assertion of Theorem 3.4.1 we have that

$$S_{N,n} \xrightarrow{d} G_N = \sum_{c=0}^{N}\frac{(\lambda\rho\gamma^2)^{c/2}}{c!}H_c(\mathcal{N}),$$

as $n \to \infty$, since for any $c = 1, 2, \ldots$

$$\frac{(\sqrt{n}m)^c}{\binom{n}{c}c!} \to \sqrt{\lambda}.$$

Let us define also

$$T_{N,n} = \sum_{c=N+1}^{m_n}\binom{m}{c}U_c^{(m,n)}.$$

Since $U_c^{(m,n)}$ are orthogonal, then by (3.15)

$$Var\left(T_{N,n}\right) = \sum_{c=N+2}^{m_n}\frac{\binom{m}{c}\gamma^{2c}}{\binom{n}{c}}\sum_{r=0}^{c}\binom{m-r}{c-r}\frac{\rho^{c-r}(1-\rho)^r}{r!}$$

The inner sum above is majorized by

$$\binom{m}{c}\sum_{r=0}^{c}\frac{1}{r!}(1-\rho)^r\rho^{c-r} \leq \binom{m}{c}e$$

since $0 < \rho \leq 1$. Thus

$$Var\left(T_{N,n}\right) \leq \sum_{c=N+1}^{m_n}\frac{\binom{m}{c}^2}{\binom{n}{c}}\gamma^{2c}\sum_{r=0}^{c}\frac{1}{r!} \leq e\sum_{c=N+1}^{\infty}\left(\frac{m^2}{n}\right)^c\frac{\gamma^{2c}}{c!}.$$

The last inequality holds true in view of

$$c! \frac{\binom{m}{c}^2}{\binom{n}{c}} \le \left(\frac{m^2}{n} \right)^c$$

which follows from the inequality

$$\frac{(m-r)^2}{n-r} \le \frac{m^2}{n}$$

valid for $0 \le r \le m \le n$. Thus for n sufficiently large

$$Var\,(T_{N,n}) \le e \sum_{c=N+1}^{\infty} \frac{(2\lambda\gamma^2)^c}{c!} = a_N$$

and $a_N \to 0$ as $N \to \infty$.

The final part of the proof follows now exactly along the lines of the proof of Theorem 3.4.2 with γ^2 replaced by $\rho\gamma^2$. This completes the case $\lambda > 0$.

For the case $\lambda = 0$, let us write

$$\frac{\sqrt{n}}{m} \left(\frac{Per\,(\mathbb{X})}{\binom{n}{m}m!\mu^m} - 1 \right) = \frac{\sqrt{n}}{m} \binom{m}{1} U_1^{(m,n)} + R_{m,n} = \frac{W_1(n)}{\sqrt{nm}} + R_{m,n},$$

where

$$R_{m,n} = \frac{\sqrt{n}}{m} \sum_{c=2}^{m} \binom{m}{c} U_c^{(m,n)}.$$

We will prove that $R_{m,n}$ converges in probability to zero. To this end note that starting from (3.15), similarly as above, we get

$$Var\,R_{m,n} = \frac{n}{m^2} \sum_{c=2}^{m} \frac{\binom{m}{c}\gamma^{2c}}{\binom{n}{c}} \sum_{r=0}^{c} \frac{1}{r!} \binom{m-r}{c-r}(1-\rho)^r \rho^{c-r}$$

$$\le e\frac{m^2}{n} \sum_{c=2}^{m} \frac{\gamma^{2c}}{c!} \le e^{1+\gamma^2} \frac{m^2}{n}.$$

Hence $Var\,R_{m,n} \to 0$, as $m^2/n \to 0$ by the assumption, and the result follows via the Tchebychev inequality.

Thus, the final result follows by the second part of Theorem 3.4.1 since we assume $m = m_n \to \infty$. □

In the hypothesis of second parts of Theorem 3.4.2 and 3.4.3 we have required the permanents to be of infinite order, that is, that $m_n \to \infty$ as $n \to \infty$. However, note that this restriction in not necessary and the (slightly modified) conclusion of Theorem 3.4.2 and 3.4.3 remains valid for m being a fixed constant. Indeed, note that the conclusions of Theorem 3.4.1 remain true also for a constant m. This follows in view of the two facts related to the proof of Theorem 3.4.1. Namely, if m is constant then the Lindeberg condition (LC) is trivially satisfied and the remainder term R_n vanishes, i.e. the properly normalized $W_1(n)$ simply equals $V_1(n)$. Consequently, the reasoning similar to the one given in the proof of Theorem 3.4.3 yields

$$\frac{\sqrt{n}}{m} \left(\frac{Per\, \mathbb{X}^{(m,n)}}{\binom{n}{m} m!\, \mu^m} - 1 \right) \xrightarrow{d} \tau \mathcal{N},$$

where $\tau^2 = (\rho + (1 - \rho)/m)\gamma^2$. Similar argument also applies to the Theorem 3.4.2. In this case $\rho = 0$ and the assertion of the theorem remains unchanged for m being a fixed constant. Note that for $m = 1$ the result is simply a classical central limit theorem (see, Corollary 4.3.1).

3.5 Additional Central Limit Theorems

In the previous section we have proved limit theorems for permanents under the column-exchangeability and column-independence assumptions (A1-A2) imposed on $\mathbb{X}^{(\infty)}$. The first of these assumptions allowed us to utilize some powerful results from the theory of exchangeable random variables like CLT and SLLN for exchangeable sequences. However, under many circumstances it is not necessary to require column-exchangeability in order to prove asymptotic results for permanents. In this section for the purpose of illustration we present two CLT for permanents without requiring the column random sequences $X^{(1)}, \ldots, X^{(m)}$ of $\mathbb{X}^{(\infty)}$ to be exchangeable. Instead we assume a weaker condition that the columns have a homogeneous correlation structure, that is for any $i_1 \neq i_2$ the correlation $corr(X_{i_1,j}, X_{i_2,j}) = \rho$ is constant. The condition (A2) remains valid, i.e. we still assume identical distributions and independence for the column random sequences $X^{(1)}, \ldots, X^{(m)}$ of $\mathbb{X}^{(\infty)}$.

Under these modified assumptions on $\mathbb{X}^{(\infty)}$ we present two versions of a permanent CLT dealing with the case when $\rho > 0$ and $\rho = 0$, respectively. As a trade-off, however, we shall need some additional hypothesis on the moments of the underlying probability law. It is important to emphasize that the formulas for the variances of $Per_h(\mathbb{X}^{(m,n)})$ derived for $\mathbb{X}^{(\infty)}$ satisfying both (A1) and (A2) still remain valid.

In Theorem 3.5.1 below, we consider the case $\rho > 0$. Indeed, a brief inspection of the proof reveals that in the case $\rho = 1$ or m being a fixed constant independent of n the result holds true as long as $0 < \sigma^2 < \infty$. Let us also note that the theorem remains valid if we assume only that $E\,|X_{11}|^{2+\delta} < \infty$ for some $0 < \delta \leq 1$ but strengthen the assumptions on the rates of m and n to $m^2/n^\delta \to 0$. It seems, however, that in general the assumption of the existence of a higher moment cannot be removed without the row-exchangeability requirement.

Theorem 3.5.1. *Assume that $E\,|X_{11}|^3 < \infty$, $\mu \neq 0$ and $\rho \in (0,1]$. If $m^2/n \to 0$ then*

$$\frac{\sqrt{n}}{m}\left(\frac{Per\,\mathbb{X}^{(m,n)}}{\binom{n}{m}\,m!\,\mu^m} - 1\right) \xrightarrow{d} \tau\mathcal{N},$$

where $\tau^2 = \rho\gamma^2$ if $m \to \infty$ and $\tau^2 = (\rho + (1-\rho)/m)\gamma^2$ for constant m.

Proof. Without any loss of generality we assume that $\mu = 1$. By (3.13) we may write

$$\frac{\sqrt{n}}{m}\left(\frac{Per\,\mathbb{X}^{(m,n)}}{\binom{n}{m}\,m!} - 1\right) = \frac{\sqrt{n}}{m}\sum_{c=1}^{m}\binom{m}{c}U_c^{(m,n)}.$$

First, note that by the argument identical to that used in the proof of Theorem 3.4.3

$$R_{m,n} = \frac{\sqrt{n}}{m}\sum_{c=2}^{m}\binom{m}{c}U_c^{(m,n)} \xrightarrow{P} 0.$$

Consequently, we need only to show that

$$\frac{\sqrt{n}}{m}\binom{m}{1}U_1^{(m,n)} = \frac{1}{m\sqrt{n}}\sum_{i=1}^{m}\sum_{j=1}^{n}\tilde{X}_{ij} = \frac{1}{\sqrt{n}}\sum_{j=1}^{n}Y_j^{(m)} \xrightarrow{d} \tau\mathcal{N},$$

where $Y_j^{(m)} = \sum_{i=1}^{m}\tilde{X}_{ij}/m$, $j = 1,\dots,n$. Let us consider an arbitrary sequence (m_n) such that $m_n^2/n \to 0$, as $n \to \infty$, and denote $Y_{nj} = Y_j^{(m_n)}/\sqrt{n}$, $j = 1,\dots,n$. Due to the structure of the matrix $\mathbb{X}^{(\infty)}$ the triangular array (Y_{nj}) is row-wise independent and identically distributed. Furthermore, the entries of the array have zero means. Hence, by CLT for the row-wise independent triangular arrays, it suffices to show that

(i) $\sum_{j=1}^{n} Var\,Y_{nj} \to C > 0$,
(ii) $\sum_{j=1}^{n} E\,Y_{nj}^2\,I(|Y_{nj}| > \varepsilon) \to 0 \quad \forall\,\varepsilon > 0$.

Note that $\sum_{j=1}^{n} Var\,Y_{nj} = (\rho + (1-\rho)/m_n)\sigma^2 \to \tau^2$ and hence (i) follows immediately. In order to check the Lindeberg condition (ii) we use Lemma 3.5.1 (see below) obtaining for any $\varepsilon > 0$

$$\sum_{j=1}^{n} E\,Y_{nj}^2\,I(|Y_{nj}| > \varepsilon)$$

$$= m_n^{-2}E\,(\tilde{X}_{11} + \dots + \tilde{X}_{m_n 1})^2 I(|\tilde{X}_{11} + \dots + \tilde{X}_{m1}| > m_n\sqrt{n}\varepsilon)$$

$$\leq \frac{1 + 4(m_n - 1)^2}{m_n}\,E\,\tilde{X}_{11}^2 I(|\tilde{X}_{11}| > \sqrt{n}\varepsilon)$$

$$\leq \frac{1 + 4(m_n - 1)^2}{m_n\sqrt{n}\varepsilon}E\,|\tilde{X}_{11}|^3 \to 0.$$

$\square$

Our next result establishes CLT for random permanents in the case when the column entries are uncorrelated. Similarly as in Theorem 3.5.1 here also one can somewhat relax the moment assumptions and require only that $E|X_{11}|^{2+\delta} < \infty$ for some $\delta > 0$, as long as it is true that $m^{2+\delta}/n^{\delta} \to 0$. It is perhaps somewhat surprising that apparently without assuming further structure of the joint distribution of the row vectors of $\mathbb{X}^{(\infty)}$ one cannot eliminate these additional assumptions.

Theorem 3.5.2. *Assume that $E X_{11}^4 < \infty$, $\mu \neq 0$, $0 < \gamma < \infty$ and $\rho = 0$. If $m^3/n \to 0$ then*

$$\sqrt{\frac{n}{m}} \left(\frac{Per\,\mathbb{X}^{(m,n)}}{\binom{n}{m} m!\, \mu^m} - 1 \right) \xrightarrow{d} \gamma \mathcal{N}.$$

Proof. As in the proof of Theorem 3.5.1 we assume, without loss of generality that $\mu = 1$. Now again by (3.13) we have

$$\sqrt{\frac{n}{m}} \left(\frac{Per\,\mathbb{X}}{\binom{n}{m} m!} - 1 \right) = \sqrt{\frac{n}{m}} \binom{m}{1} U_1^{(m,n)} + R_{m,n},$$

where

$$R_{m,n} = \sqrt{\frac{n}{m}} \sum_{c=2}^{m} \binom{m}{c} U_c^{(m,n)}.$$

We show first that $R_{m,n}$ tends to zero in probability. To this end observe that by (3.14), (3.15) and the fact that $\binom{m}{c}/\binom{n}{c} \leq (m/n)^c$, $(1 \leq c \leq m \leq n)$, it follows that (taking n large enough to have $m/n \leq 1$)

$$Var\,R_{m,n} = \frac{n}{m} \sum_{c=2}^{m} \frac{\binom{m}{c} \gamma^{2c}}{\binom{n}{c} c!} \leq \frac{m}{n} \exp(\gamma^2) \to 0,$$

and hence $R_{m,n} \to 0$ in probability, in view of the Tchebychev inequality.

Secondly, we will show that

$$\sqrt{\frac{n}{m}} \binom{m}{1} U_1^{(m,n)} = \frac{1}{\sqrt{mn}} \sum_{i=1}^{m} \sum_{j=1}^{n} \tilde{X}_{ij} = \frac{1}{\sqrt{n}} \sum_{j=1}^{n} Y_j^{(m)} \xrightarrow{d} \gamma \mathcal{N}.$$

where $Y_j^{(m)} = \sum_{i=1}^{m} \tilde{X}_{ij}/\sqrt{m}$, $j = 1, \ldots, n$. As in the proof of Theorem 3.5.1 let us again consider an arbitrary sequence (m_n) such that $m_n^3/n \to 0$, as $n \to \infty$, and denote $Y_{nj} = Y_j^{(m_n)}/\sqrt{n}$, $j = 1, \ldots, n$. Again, due to the structure of the matrix $\mathbb{X}$ the triangular array (Y_{nj}) is row-wise independent and identically distributed. Furthermore, the entries of the array have zero means. Hence to complete the proof we may again use the CLT for rowwise independent triangular arrays. Since now $\sum_{j=1}^{n} Var\, Y_{nj} = \gamma^2$ we need only to verify the Lindeberg condition. By Lemma 3.5.1 below it follows that for any $\varepsilon > 0$

$$\sum_{j=1}^{n} E Y_{nj}^2 I(|Y_{nj}| > \varepsilon)$$

$$= m_n^{-1} E (\tilde{X}_{11} + \ldots + \tilde{X}_{m_n 1})^2 I(|\tilde{X}_{11} + \ldots + \tilde{X}_{m_n 1}| > \sqrt{m_n n}\, \varepsilon)$$

$$\leq (1 + 4(m_n - 1)^2)\, E \tilde{X}_{11}^2 I(|\tilde{X}_{11}| > \sqrt{n/m_n}\, \varepsilon)$$

$$\leq \frac{(1 + 4(m_n - 1)^2) m_n}{n}\, E |\tilde{X}_{11}|^4 \to 0$$

since $m^3/n \to 0$. The proof is thus completed as soon as the result of Lemma 3.5.1 is justified. $\qquad \square$

Lemma 3.5.1. *Let $(X_1, \ldots, X_n)$ be a random vector with square integrable, identically distributed components. Then for any $a > 0$*

$$E (X_1 + \ldots + X_n)^2 I(|X_1 + \ldots + X_n| > a) \leq n(1 + 4(n-1)^2) E X_1^2 I(|X_1| > a/n).$$

Proof. Observe that

$$E (X_1 + \ldots + X_n)^2 I(|X_1 + \ldots + X_n| > a)$$

$$\leq \sum_{i=1}^{n} E (X_1 + \ldots + X_n)^2 I(|X_i| > a/n)$$

$$= \sum_{i=1}^{n} \left(E X_i^2 I(|X_i| > a/n) + \sum_{j \neq i}^{n} E X_j^2 I(|X_i| > a/n) \right.$$

$$\left. + \sum_{j \neq i}^{n} E X_i X_j I(|X_i| > a/n) + \sum_{\substack{1 \leq j < k \leq n \\ j \neq i \neq k}} E X_j X_k I(|X_i| > a/n) \right).$$

$$(3.19)$$

But for any X, Y, Z identically distributed, square integrable random variables and any positive b we have

$$E X^2 I(|Z| > b)$$

$$= E X^2 I(|X| > b) I(|Z| > b) + E X^2 I(|X| \leq b) I(|Z| > b)$$

$$\leq E X^2 I(|X| > b) + E Z^2 I(|Z| > b)$$

$$= 2 E X^2 I(|X| > b),$$

Similarly, but additionally using the Cauchy-Schwartz, inequality we get

$$E |XZ| I(|Z| > b)$$

$$= E |XZ| I(|X| > b) I(|Z| > b) + E |XZ| I(|X| \leq b) I(|Z| > b)$$

$$\leq \sqrt{E X^2 I(|X| > b) E Z^2 I(|Z| > b)} + E Z^2 I(|Z| > b)$$

$$= 2 E X^2 I(|X| > b).$$

Now, by the above inequality, it follows also that

$$
\begin{aligned}
& E\,|XY|\,I(|Z| > b) \\
& = E\,|XY|\,I(|X| > b)I(|Z| > b) + E\,|XY|\,I(|X| \le b)I(|Z| > b) \\
& \le E\,|XY|\,I(|X| > b) + E\,|ZY|\,I(|Z| > b) \\
& \le 4\,E\,X^2 I(|X| > b).
\end{aligned}
$$

Finally, applying the above three inequalities to the right hand side of (3.19) we get

$$
\begin{aligned}
& E\,(X_1 + \ldots + X_n)^2\,I(|X_1 + \ldots + X_n| > a) \\
& \le n[E\,X_1^2\,I(|X_1| > a/n) + 2(n-1)E\,X_1^2\,I(|X_1| > a/n) \\
& \quad + 2(n-1)E\,X_1^2\,I(|X_1| > a/n) + 4(n-1)(n-2)E\,X_1^2\,I(|X_1| > a/n)] \\
& = n(1 + 4(n-1)^2)E\,X_1^2\,I(|X_1| > a/n)
\end{aligned}
$$

which gives the required inequality. $\qquad\square$

3.6 Strong Laws of Large Numbers

By virtue of the decomposition (3.13) derived in Chapter 2 we may represent a random permanent as a sums of uncorrelated backward martingales . This is helpful for analyzing asymptotic behavior, as it turns out that the reverse martingales have particularly nice convergence properties. For instance, for the discrete martingales $(T = 0, 1, \ldots)$ we have the following.

Theorem 3.6.1. *Let $(Y_t)_{t=0,1,\ldots}$ be a backward martingale adopted to a decreasing sequence $(\mathcal{F}_t)_{t=0,1,\ldots}$ of σ-fields (sub-σ-fields of $\mathcal{F}$) and let $\mathcal{F}_\infty = \bigcap_{t=0}^{\infty} \mathcal{F}_t$. Then Y_t converges almost surely and in L_1 to $E(Y_1|\mathcal{F}_\infty)$.* $\qquad\square$

The proof of this theorem may be found in any standard general textbook on martingale theory (see, e.g. Chow and Teicher, 1978, chapter 7). As an example of its applicability let us consider the following.

Lemma 3.6.1 (SLLN for classical U-statistics). *Let k be a given positive integer. Consider a sequence of U-statistics $U_l^{(k)}(h)$, $l = k, k+1, \ldots$, with the fixed kernel h. If $E|h| < \infty$ then*

$$
U_l^{(k)}(h) \to Eh \quad a.s. \quad as\ l \to \infty.
$$

Proof. We shall use the H-decomposition (1.15). Since k is fixed the result will follow if we argue that $U_{c,l} \to 0$ as $l \to \infty$ for each $c = 1, \ldots, k$. To this end note that it follows from Theorem 2.2.2 (for $\mathbb{X} = \underline{X}$, $m_n \equiv k$ and $n = l$) that for each $c = 1, \ldots, k$ the variables $U_{c,l}, l = c, c+1, \ldots$ form backward martingale sequence with respect to the σ-field $\mathcal{F}_{c,l} = \sigma\{U_{c,l}, U_{c,l+1}, \ldots\}$.

Thus there exists a random variable X_c which is measurable with respect to the σ-field $\mathcal{F}_\infty = \bigcap_{s=c}^\infty \mathcal{F}_{c,s}$ such that as $l \to \infty$

$$U_{c,l} \to X_c \qquad a.s. \quad \text{and} \quad EU_{c,l} \to EX_c \qquad c = 1 \ldots, k.$$

However, by the Hewitt-Savage zero-one law (cf. e.g., Chow and Teicher 1978) the tail σ-field $\mathcal{F}_\infty$ is trivial and thus X_c has to be a constant a.s. This, however, in view of the above convergence of the expectations implies that for $c = 1, \ldots, k$ we have necessarily $X_c = EX_c = 0$ since $EU_{c,l} = 0$ for all $l \geq k \geq c$. □

As a simple example of the above consider the law of large numbers for a sum of independent and identically distributed random variables.

Example 3.6.1 (Classical SLLN). Let (X_i) be a sequence of independent and identically distributed, integrable random variables and let $S_l = \sum_{i=1}^l X_i$. Then

$$S_l/l \to EX_1 \quad a.s. \quad as \quad l \to \infty.$$

In general for $Per\,\mathbb{X}^{(m,n)}$ when both $m, n \to \infty$ the results of the type presented above are insufficient since the number of elements in the H-decomposition increases with n and thus some refinements of the result of Lemma 3.6.1 are needed. The first results treating almost sure convergence of permanents for one dimensional projection matrices, that is the issue of strong convergence for random permanents were obtained in Halász and Székely (1976). The authors considered symmetric polynomials of increasing order for positive iid random variables which, for the matrix of positive entries, is equivalent to taking $\rho = 1$ in the scheme (A1-A2) we consider here. Then it was shown that under the condition $m/n \to \lambda \geq 0$

$$\left(\frac{Per\,\underline{\mathbb{X}}}{\binom{n}{m}} \right)^{1/m} \to \mathcal{S}(\lambda) \qquad a.s.,$$

where $\mathcal{S}(\lambda)$ is a non-random quantity uniquely determined by the value of λ and such that $\mathcal{S}(0) = \mu$. Note that the above result is implied by the relation of the type

$$\frac{Per\,\mathbb{X}^{(m,n)}}{\binom{n}{m}m!\mu^m} \to 1 \quad a.s. \tag{3.20}$$

as $n \to \infty$ and/or $m \to \infty$, but only when $\lambda = 0$. Further non-linear laws of large numbers for permanents are also known. In particular, it is known that for $\mathbb{X} = \underline{\mathbb{X}}$ the relation (3.20) cannot hold if $m/\sqrt{n} \to \lambda > 0$ and only a logarithmic version of SLLN for elementary symmetric polynomials under restrictive technical assumptions were obtained.

Here we discuss the conditions under which the SLLN of the form (3.20) holds under the exchangeability assumptions (A1-A2) on $\mathbb{X}^{(\infty)}$ as given in Chapter 1. It turns out that in this setting a somewhat more stringent condition on the rate of relative asymptotic behavior of m and n is helpful.

Theorem 3.6.2. *Let $m = m_n$ be a non-decreasing sequence. If $\rho \in (0,1]$ and $m^p/n \to 0$ for some $p > 2$ then*

$$\frac{Per\, \mathbb{X}^{(m,n)}}{\binom{n}{m} m! \mu^m} \to 1 \quad a.s.$$

Proof. As before we use the H-decomposition (3.13). Hence, it suffices to prove that

$$\sum_{c=1}^{m} \binom{m}{c} U_c^{(m,n)} \to 0 \quad a.s.$$

as $n \to \infty$.

Observe that for sufficiently large n, say $n > n_0$, we have $m = m_n \leq n^{1/p}$. First, we prove that for any arbitrary but fixed $c \geq 1$

$$\binom{m}{c} U_c^{(m,n)} \to 0 \quad a.s.$$

To this end note that for any $\varepsilon > 0$

$$P\left(\sup_{n \geq 2^{n_0}} \binom{m_n}{c} |U_c^{(m,n)}| > \varepsilon\right)$$

$$\leq \sum_{k \geq n_0} P\left(\max_{2^k \leq n \leq 2^{k+1}} \binom{m_n}{c} |U_c^{(m,n)}| > \varepsilon\right).$$

Since

$$\binom{m_n}{c} \leq \frac{m_n^c}{c!} \leq \frac{n^{c/p}}{c!},$$

it follows that

$$P\left(\sup_{n \geq 2^{n_0}} \binom{m_n}{c} |U_c^{(m,n)}| > \varepsilon\right)$$

$$\leq \sum_{k \geq n_0} P\left(\frac{2^{c(k+1)/p}}{c!} \max_{2^k \leq n \leq 2^{k+1}} |U_c^{(m,n)}| > \varepsilon\right).$$

Now, by the maximal inequality for backward martingales (see Theorem 2.1.2) we get

$$P\left(\sup_{n \geq 2^{n_0}} \binom{m_n}{c} |U_c^{(m,n)}| > \varepsilon\right) \leq \sum_{k \geq n_0} \frac{Var\left(U_c^{(m_{2^k}, 2^k)}\right) 2^{2c(k+1)/p}}{(c!)^2 \varepsilon^2}$$

$$\leq \sum_{k \geq n_0} \frac{(\gamma^2 \rho)^c e}{(c!)^2 \varepsilon^2} \frac{2^{2c(k+1)/p}}{\binom{2^k}{c}},$$

since it follows (see Example 2.3.1) for $\rho > 0$ that

$$Var\, U_c^{(m,n)} \leq \frac{(\gamma^2 \rho)^c e}{\binom{n}{c}}.$$

Let us note that for sufficiently large n we have $\binom{n}{c} \geq \frac{n^c}{2^c c!}$ which entails

$$P\left(\sup_{n \geq 2^{n_0}} \binom{m_n}{c} |U_c^{(m,n)}| > \varepsilon\right)$$

$$\leq \frac{(2\gamma^2 \rho)^c e}{c! \varepsilon^2} \sum_{k \geq n_0} \frac{2^{2c(k+1)/p}}{2^{kc}} = \frac{(2^{1+1/p} \gamma^2 \rho)^c e}{c! \varepsilon^2} \sum_{k \geq n_0} 2^{ck(2/p-1)}$$

and the last series converges since $p > 2$, i.e. $2/p - 1 < 0$. Thus, it follows that

$$\lim_{n_0 \to \infty} P\left(\sup_{n \geq 2^{n_0}} \binom{m_n}{c} |U_c^{(m,n)}| > \varepsilon\right) = 0,$$

and, consequently, $\binom{m_n}{c} |U_c^{(m,n)}| \to 0$ a.s.

In the second step we consider

$$R_{m,n}(k) = \sum_{c=k}^{m} \binom{m}{c} U_c^{(m,n)}$$

for some $2 \leq k \leq m$. We will prove that there exists k satisfying $2 \leq k \leq m$ such that $R_{m,n}(k)$ converges to zero completely, i.e., $\forall\, \varepsilon > 0$

$$\sum_{n=1}^{\infty} P(|R_{m,n}(k)| > \varepsilon) < \infty.$$

This will obviously follow if we can show that

$$\sum_{n=1}^{\infty} Var(R_{m,n}(k)) < \infty.$$

By the orthogonality of the H-decomposition (3.14) it follows that

$$Var(R_{m,n}(k)) = \sum_{c=k}^{m} \binom{m}{c}^2 Var(U_c^{(m,n)}).$$

Consequently, using (3.15) and proceeding similarly as in previous proofs we get

$$Var(R_{m,n}(k)) = \sum_{c=k}^{m} \frac{\binom{m}{c}}{\binom{n}{c}} \gamma^{2c} \sum_{r=0}^{c} \binom{m-r}{c-r} \frac{\rho^{c-r}(1-\rho)^r}{r!}$$

$$\leq e \left(\frac{m^2}{n}\right)^k \sum_{c=k}^{m} \left(\frac{m^2}{n}\right)^{c-k} \frac{\gamma^{2c}}{c!}.$$

Now, for large enough n we have $m^2/n < m^p/n < 1$, and hence

$$Var(R_{m,n}(k)) \le \left(\frac{m^2}{n}\right)^k e^{1+\gamma^2}.$$

Since

$$\frac{m^2}{n} = \left(\frac{m^p}{n}\right)^{2/p} \frac{1}{n^{1-2/p}} \le \frac{1}{n^{1-2/p}}$$

for sufficiently large n, it follows that for k such that $k(1 - 2/p) > 1$ the sequence $(R_{m,n}(k))$ converges completely to zero as $n \to \infty$.

Combining the two steps we arrive at the conclusion of the theorem. $\square$

Let us now turn attention to the case $\rho = 0$. In case of the permanent SLLN for uncorrelated (and thus possibly independent) within-column components, an additional technical condition on the behavior of the sequence $m = m_n$ is required.

Theorem 3.6.3. *Let $\rho = 0$ and let $m = m_n$ be a non-decreasing sequence. Assume that $\exists\, p > 1$ and $\exists\, L > 0$ such that*

$$m^p/n \to 0$$

and

$$m_{2n}^2 < L m_n n^{1/p}. \tag{3.21}$$

Then

$$\frac{Per\, \mathbb{X}^{(m,n)}}{\binom{n}{m} m! \mu^m} \to 1, \quad a.s.$$

Proof. Observe that in the case $\rho = 0$ it follows from (3.15) that

$$Var(U_c^{(m,n)}) = \frac{\gamma^{2c}}{c!\binom{m}{c}\binom{n}{c}}.$$

Consequently, using the same argument as in the previous proof we arrive at

$$P\left(\sup_{n\ge 2^{n_0}} \binom{m}{c}|U_c^{(m,n)}| > \varepsilon\right) \le \sum_{k\ge n_0} \frac{m_{2^{k+1}}^{2c}}{c!^2\varepsilon^2} Var\left(U_c^{(m_{2^k},2^k)}\right)$$

$$\le \frac{(2\gamma)^{2c}}{c!\varepsilon^2} \sum_{k\ge n_0} \left(\frac{m_{2^{k+1}}^2}{m_{2^k}2^k}\right)^c,$$

where we used the inequalities $s(s-1)\dots(s-c+1) \ge (s/2)^c$ for s sufficiently large, taking $s = m_{2^k}$ and $s = 2^k$.

Now by (3.21) it follows that

$$P\left(\sup_{n\ge 2^{n_0}} \binom{m}{c}|U_c^{(m,n)}| > \varepsilon\right) \le \frac{L(2\gamma)^{2c}}{c!\varepsilon^2} \sum_{k\ge n_0} 2^{kc(1/p-1)}$$

and the series converges, since $p > 1$. Hence for any fixed c it follows that $\binom{m}{c}|U_c^{(m,n)}| \to 0$ a.s.

Similarly as in the proof of Theorem 3.6.2 we again consider $R_{m,n}(k) = \sum_{c=k}^{m} \binom{m}{c} U_c^{(m,n)}$ for some $2 \le k \le m$. We shall prove that there exists k $(2 \le k \le m)$ such that $R_{m,n}(k)$ converges to zero completely.

To this end it suffices to note that

$$Var(R_{m,n}(k)) = \sum_{c=k}^{m} \frac{\binom{m}{c}}{\binom{n}{c}} \frac{\gamma^{2c}}{c!} \le \sum_{c=k}^{m} \frac{m^c 2^c \gamma^{2c}}{n^c c!}$$

$$= \left(\frac{m}{n}\right)^k \sum_{c=k}^{m} \left(\frac{m}{n}\right)^{c-k} \frac{(2\gamma^2)^c}{c!} \le \left(\frac{m}{n}\right)^k e^{2\gamma^2}.$$

But now it follows immediately that the series of variances converges if only the parameter k is chosen in such a way that $k(1 - 1/p) > 1$, since

$$\frac{m}{n} = \left(\frac{m^p}{n}\right)^{1/p} \frac{1}{n^{1-1/p}} \le \frac{1}{n^{1-1/p}}$$

for n sufficiently large to have $m^p/n < 1$. □

To illustrate the application of this result let us revisit the by now familiar example of number of perfect matchings in random graphs.

Example 3.6.2 (SLLN for perfect matchings). Consider as before the bipartite graph $G = (V_1, V_2; E)$ with the matrix $\mathbb{X}^{(m,n)}$ of independent and identically distributed Bernoulli weights (i.e., the edges occur independently with a fixed probability $0 < p < 1$). We are again interested in the asymptotic behavior, this time in the almost sure sense, of the number of perfect matchings $\mathcal{H}(G, \mathbb{X}^{(m,n)})$. Under the assumptions of Theorem 3.6.3 we have

$$\frac{\mathcal{H}(G, \mathbb{X}^{(m,n)})}{\binom{n}{m} m! p^m} \to 1 \qquad a.s.$$

The condition (3.21) is somewhat mysterious, but in essence seems to be a requirement for the behavior of the sequence (m_n). Observe for instance that if m_{2n}/m_n is bounded then the condition (3.21) holds.

Note also that instead of (3.21) one could require that the sequence m_n satisfies

$$\sum_{n=1}^{\infty} \frac{1}{m_n n^{2-2/p}} < \infty.$$

To see this, note

$$P\left(\sup_{n \ge 2^{n_0}} \binom{m_n}{c} |U_c^{(m,n)}| > \varepsilon\right) \le \frac{2^{2/p}(2\gamma)^{2c}}{c!\varepsilon^2} \sum_{k \ge n_0} \frac{2^k}{m_{2^k} 2^{k(2-2/p)}},$$

which follows from the inequality $m_n < n^{1/p}$ holding true for sufficiently large n. The fact that the last series converges follows in view of the Cauchy condensation criterion and (3.6).

The above implies in particular that if $p > 2$ the assumption (3.21) may be dropped, since the condition (3.6) is then always satisfied.

3.7 Bibliographic Details

A very good introduction to the limiting theory for exchangeable random variables (including the De Finetti-type theorems) is given in Taylor et al. (1985) and Chow and Teicher (1978). A deep treatment of the subject of exchangeability is also offered in Aldous (1985). The results on elementary symmetric polynomials (i.e., permanent of one dimensional projection matrices) of Theorem 3.1.1 are discussed in the papers of Székely (1982), van Es and Helmers (1988) Borovskikh and Korolyuk (1994) as well as Rempała and Wesołowski (1999).

The proof of limit theorem for permanents as given in Theorem 3.3.1 as well as some results on laws of large numbers for triangular arrays are taken from Rempała and Wesołowski (2004). Some of the discussion related to the elementary symmetric and Hermité polynomials is taken from Avram and Taqqu (1986) (including the material on the Waring function). The majority of the material in Section 3.5 comes from Rempała and Wesołowski (2002b). Some results of the similar flavor may be found in the works of Korolyuk and Borovskikh (1992), Korolyuk and Borovskikh (1995), Janson (1994) and Kaneva and Korolyuk (1996) as well as in the monograph by Girko (1990). An interested reader should refer to those works for more details. The applications of some of these asymptotic results to statistical estimation and to inverse moments expansion problems are discussed in Rempała and Székely (1998) and Rempała (2004).

The results on the strong convergence properties of elementary symmetric polynomials are treated in Móri and Székely (1982) where the saddle point approximation method is used instead of the H-decomposition to obtain the limiting results.

Some further results on the strong convergence of P-statistics with general kernels are discussed in Rempała and Gupta (2000). The proofs therein also appeal directly to the martingale convergence theorems rather then the use of the H-decomposition. The results stated here as Theorems 3.6.2 and 3.6.3 come from the paper of Rempała and Wesołowski (2002c). Some extensions are also discussed in Fyodorov (2006).

4

Weak Convergence of Permanent Processes

4.1 Introduction

In the previous chapter we have discussed two results on the weak convergence of random permanents in Theorems 3.4.2 and 3.4.3. The results were proved, in essence, by reducing the analysis of asymptotic behavior of a random permanent to that of an elementary symmetric polynomial and taking advantage of the general limit theorem for elementary symmetric polynomials of increasing order (cf. Theorem 3.3.1). Whereas the advantage of the approach is its conceptual simplicity, we saw that it required overcoming several technical difficulties.

One way of avoiding some of these difficulties is by utilizing the results borrowed from a quite general theory on the weak convergence of the stochastic integrals developed by Jakubowski et al. (1989) and Kurtz and Protter (1991). By adopting this approach, one may obtain a functional analogue of Theorem 3.3.1, for instance, as a consequence of the result on weak convergence of sequences of stochastic integrals given in Kurtz and Protter (1991). Then, similarly as in Chapter 3, one may relate that result to a functional version of the limit theorem for random permanents. The advantage of applying this method is in considerable simplification of the arguments leading to analogues of Theorems 3.4.2 and 3.4.3 which in current setting are obtained as simple corollaries of the appropriate functional limit theorems. The disadvantage is, of course, that one needs to introduce some new tools from the modern theory of stochastic integration. It is perhaps useful to note that, unlike in Chapter 5 where the stochastic integrals shall be used to write out the appropriate weak limits in more concise form, in the present chapter we utilize the stochastic integration theory mostly for the sake of the suitable convergence properties of the processes of interest which turn out to be stochastic integrals with respect to certain continuous martingales.

The developments presented in the current chapter, though leading to results for permanent processes, may be also viewed as an alternative way of arriving at the limit theorems for random permanents given in Chapter 3.

The current chapter is intended to be somewhat self-contained in that it does not utilize any of the results on permanents derived in the previous chapter. In the next two sections we recall some basic properties of the space $D_{\mathbf{R}^k}$ which is the Skorohod space of cadlag functions on $[0, 1]$ with values in $\mathbf{R}^k$. In the subsequent Section 4.4 we define a stochastic process which is related to a permanent, the co-called *permanent stochastic process* or PSP. Again, similarly as in Chapter 3, the orthogonal decomposition of a permanent is our major tool also in this chapter. In Section 4.5 we recall some basic facts from the stochastic integration theory and then derive an invariance principle for what we refer to as an elementary symmetric polynomial process (ESPP) via the stochastic integrals convergence theorem. The result is presented as Theorem 4.5.1. Subsequently, the result on the ESPP is related to the asymptotic behavior of the component processes of PSP (Theorem 4.6.1). Once this result is established the asymptotics for PSP follow via the truncation-type argument along with the basic properties of the Prohorov distance and the martingale properties of the component processes. These are the main results of this chapter and are presented in the last two theorems of Section 4.7. In particular, from these theorems the assertions of Theorems 3.4.2 and 3.4.3 follow immediately.

4.2 Weak Convergence in Metric Spaces

First, we give some general background on the Skorohod spaces and the weak convergence of stochastic processes.

Let $(\mathcal{I}, d)$ be an arbitrary metric space with a metric d and let $\mathcal{B}(\mathcal{I})$ be the corresponding Borel σ-field. Let $\mathcal{P}(\mathcal{I})$ be a class of Borel probability measures on $\mathcal{I}$. So far we have been mostly concerned with the case of $\mathcal{I} = \mathbf{R}^k$. Since we shall now venture into more abstract metric spaces at this point it is perhaps useful to note that the notion of the weak convergence used so far for $\mathbf{R}^k$ also works in a more general metric space setting. Recall the following.

Definition 4.2.1 (Weak convergence in metric spaces). *Consider a sequence of probability measures (P_n) and a probability measure P on $(\mathcal{I}, \mathcal{B}(\mathcal{I}))$. We say that the sequence (P_n) converges weakly to P if and only if the convergence (recall the notational convention (1.5))*

$$P_n[f] \to P[f] \quad n \to \infty$$

holds for every bounded continuous real valued function f on $\mathcal{I}$.

In the sequel, if Y_n and Y are random elements in $\mathcal{I}$, we continue to write $Y_n \overset{d}{\to} Y$ for the weak convergence of the corresponding probability distributions. The following result ensures that the above notion of the weak convergence is consistent with the other ones used so far. Recall that a set $A \in \mathcal{B}(\mathcal{I})$ whose boundary ∂A satisfies $P(\partial A) = 0$ is called a *P-continuity set* (note that the set ∂A is closed and hence is an element of $\mathcal{B}(\mathcal{I})$).

Theorem 4.2.1 (Portmanteau's theorem). *Let P_n and P be probability measures on $(\mathcal{I}, d)$. Then the following conditions are equivalent*

(i) P_n converges weakly to P as $n \to \infty$.
(ii) $P_n f \to P f$ as $n \to \infty$ for all bounded, uniformly continuous f.
(iii) $\limsup_n P_n(F) \leq P(F)$ for all closed $F \subseteq \mathcal{I}$.
(iv) $\liminf_n P_n(G) \leq P(G)$ for all open $G \subseteq \mathcal{I}$.
(v) $P_n(A) \to P(A)$ as $n \to \infty$ for all P-continuity sets $A \in \mathcal{B}(\mathcal{I})$.

$\square$

The proof of this result may be found, for instance, in Billingsley (1999, p. 16).

It is often convenient to topologize the space of all probability measures $\mathcal{P}(\mathcal{I})$ in a way which agrees with a notion of weak convergence. This may be done for instance by introducing the *Prohorov metric*.

Definition 4.2.2 (Prohorov metric). *Let P, Q be two probability measures belonging to $\mathcal{P}(\mathcal{I})$. The Prohorov metric (or distance) is defined as*

$$\varrho(P, Q) = \inf\{\varepsilon > 0 : P(F) \leq Q(F^\varepsilon) + \varepsilon \quad \forall\, F \in \mathcal{C}\},$$

where $\mathcal{C}$ is the collection of closed subsets of $\mathcal{I}$ and

$$F^\varepsilon = \left\{ x \in \mathcal{I} : \inf_{y \in F} d(x, y) < \varepsilon \right\}.$$

In order to argue that ϱ is indeed a metric we need the following.

Lemma 4.2.1. *Let $P, Q \in \mathcal{P}(\mathcal{I})$ and $\alpha, \beta > 0$. If*

$$P(F) \leq Q(F^\alpha) + \beta \tag{4.1}$$

for all $F \in \mathcal{C}$, then

$$Q(F) \leq P(F^\alpha) + \beta \tag{4.2}$$

for all $F \in \mathcal{C}$.

Proof. Given $F_1 \in \mathcal{C}$ let $F_2 = \mathcal{I} - F_1^\alpha$ and note that $F_2 \in \mathcal{C}$ and $F_1 \subset S - F_2^\alpha$. Consequently, by (4.1) with $F = F_2$

$$P(F_1^\alpha) = 1 - P(F_2) \geq 1 - Q(F_2^\alpha) - \beta \geq Q(F_1) - \beta,$$

which implies (4.2) with $F = F_1$. $\square$

From the above lemma it follows that $\varrho(P, Q) = \varrho(Q, P)$ for all $P, Q \in \mathcal{P}(\mathcal{I})$. Also, if $\varrho(P, Q) = 0$, then $P(F) = Q(F)$ for all $F \in \mathcal{P}(\mathcal{I})$ and hence for all $F \in \mathcal{B}(\mathcal{I})$. Thus $\varrho(P, Q) = 0$ if and only if $P = Q$. For $P, Q, R \in \mathcal{P}(\mathcal{I})$, take any $\varepsilon > 0$ and $\delta > 0$ such that $\varrho(P, Q) < \varepsilon$ and $\varrho(Q, R) < \delta$. For any set A denote by $\overline{A}$ its closure. Then on noting that $\left(\overline{F^\delta}\right)^\varepsilon \subset F^{\delta + \varepsilon}$ we get

$$P(F) \leq Q\left(F^\delta\right) + \delta \leq Q\left(\overline{F^\delta}\right) + \delta \leq R\left(\left(\overline{F^\delta}\right)^\varepsilon\right) + \varepsilon + \delta \leq R\left(F^{\varepsilon + \delta}\right) + \delta + \varepsilon$$

for all $F \in \mathcal{P}(\mathcal{I})$. Hence $\varrho(P, R) \leq \delta + \varepsilon$ and due to the fact that ε and δ are arbitrary numbers satisfying $\varrho(P, Q) < \varepsilon$ and $\varrho(Q, R) < \delta$ we get that $\varrho(P, R) \leq \varrho(P, Q) + \varrho(Q, R)$ which ends the proof of the fact that ϱ is a metric.

Let X, Y be two random elements in $\mathcal{I}$ with their corresponding probability laws P and Q. The Prohorov distance between X, Y is then defined as $\varrho(X, Y) = \varrho(P, Q)$.

It turns out that if $(\mathcal{I}, d)$ is a separable metric space then the convergence in Prohorov metric is equivalent to the weak convergence as defined above. Let us state this as follows.

Theorem 4.2.2. *Let $(\mathcal{I}, d)$ be a separable metric space and let Y_n, Y be $\mathcal{I}$-valued random elements. Then, as $n \to \infty$,*

$$Y_n \xrightarrow{d} Y \quad \text{if and only if} \quad \varrho(Y_n, Y) \to 0.$$

Proof. Let $P_n, P \in \mathcal{P}(\mathcal{I})$ denote the laws of Y_n, Y respectively. Suppose that $\varrho(Y_n, Y) \to 0$, then there exists a sequence ε_n such that $\varrho(Y_n, Y) < \varepsilon_n \to 0$. Note that for any $F \in \mathcal{C}$ $\limsup_n P_n(F) \leq \limsup_n P_n(F^{\varepsilon_n}) + \varepsilon_n = P(F)$ and hence $Y_n \xrightarrow{d} Y$ by Theorem 4.2.1.

Conversely, suppose now that $Y_n \xrightarrow{d} Y$. Let $\varepsilon > 0$ be taken arbitrarily. Consider a countable family of open disjoint subsets $\{A_i\}_{i \geq 1}$ of the space $\mathcal{I}$ such that $\sup_{x, y \in A_i} d(x, y) < \varepsilon$ for all i's and $P(A_0) \leq \varepsilon/2$ for $A_0 = \mathcal{I} \setminus \bigcup_i A_i$. Such a partition of $\mathcal{I}$ exists due to the assumption of separability. Let us choose k large enough so that $P\left(\bigcup_{i > k} A_i\right) < \varepsilon/2$ and let $\mathbf{G}_0$ be a finite class of open sets of the form $A_{i_1} \cup \cdots \cup A_{i_l}$ for $1 \leq i_1 <, \ldots, < i_l \leq k$. If P_n converges weakly to P then by Theorem 4.2.1 there exist $n_0 = n_0(\varepsilon)$ such that for all $n \geq n_0$, $P_n(G) > P(G) - \varepsilon$ for each $G \in \mathbf{G}_0$. For a given set B let B_0 be the union of those sets among $\{A_i\}$ that intersect with B. Note that $B_0^\varepsilon \in \mathbf{G}_0$, and for $n \geq n_0$ we have

$$P(B) \leq P(B_0) + P\left(\bigcup_{i > k} A_i\right) + P(A_0) \leq P(B_0^\varepsilon) + \varepsilon < P_n(B_0^\varepsilon) + 2\varepsilon$$
$$\leq P_n(B^{2\varepsilon}) + 2\varepsilon,$$

and hence $\varrho(Y_n, Y) \leq \varepsilon$. $\qquad \square$

The following upper bound on the Prohorov distance shall be useful.

Theorem 4.2.3. *Let $P, Q \in \mathcal{P}(\mathcal{I})$ and let ν be a measure belonging to $\mathcal{P}(\mathcal{I}^2)$ such that $\nu(A \times \mathcal{I}) = P(A)$ and $\nu(\mathcal{I} \times A) = Q(A)$ for all $A \in \mathcal{B}(\mathcal{I})$ (i.e., ν has marginals P, Q). Then*

$$\varrho(P, Q) \leq \inf\{\varepsilon > 0 : \nu\{(x, y) : d(x, y) \geq \varepsilon\} \leq \varepsilon\}.$$

Proof. Suppose that

$$\nu\{(x,y) : d(x,y) \geq \varepsilon\} \leq \varepsilon,$$

then for any $F \in \mathcal{C}$

$$P(F) = \nu(F \times \mathcal{I})$$
$$\leq \nu((F \times \mathcal{I}) \cap \{(x,y) : d(x,y) < \varepsilon\}) + \varepsilon$$
$$\leq \nu(\mathcal{I} \times F^{\varepsilon}) + \varepsilon = Q(F^{\varepsilon}) + \varepsilon$$

and hence $\varrho(P,Q) \leq \varepsilon$ and the result follows. □

4.3 The Skorohod Space

In this section we introduce an important metric space that shall be essential for further discussions of this chapter. Let us begin with a description of a class of functions which plays a fundamental role in describing the functional limits theorems we are interested in.

Consider a space of real valued functions x on the interval $[0,1]$ that are right-continuous and have left-hand limits. More precisely,

(i) for $0 \leq t < 1$, $x(t^+) = \lim_{s \downarrow t} x(s)$ exists and $x(t^+) = x(t)$, and
(ii) for $0 < t \leq 1$, $x(t^-) = \lim_{s \uparrow t} x(s)$ exists.

Functions with these properties are typically called *cadlag* which is an acronym for French "continu à droite, limites à gauche". A function x is said to have a discontinuity of the first kind at t if $x(t^-)$ and $x(t^+)$ exist but differ. Any discontinuities of a cadlag function are of the first kind. Note also that the space of all continuous functions on $[0,1]$ is a subset of the collection of all cadlag functions on $[0,1]$.

In the space of continuous functions on $[0,1]$ one typically defines the distance between two elements x, y by considering the uniform perturbation of the difference of the graphs $x(t)$ and $y(t)$ with the same abscissas (time), that is

$$d_C(x,y) = \sup_{t \in [0,1]} |x(t) - y(t)|. \tag{4.3}$$

In contrast, in the space of cadlag functions it is convenient to additionally allow for a small deformation of the time scale. Physically, this reflects an acknowledgment that we cannot measure time with perfect accuracy more than we can measure position. The following distance notion, devised by Skorohod, embodies the idea.

Let Υ denote a class of strictly increasing, continuous functions (or mappings) from $[0,1]$ onto itself. If $v \in \Upsilon$ then $v(0) = 1$ and $v(1) = 1$. For cadlag

x and y define their distance $d(x, y)$ to be the infimum of those positive ε for which there exists $v \in \Upsilon$ satisfying

$$\sup_t |v(t) - t| = \sup_t |t - v^{-1}(t)| < \varepsilon$$

and

$$\sup_t |x(t) - y(v(t))| = \sup_t |x(v^{-1}(t)) - y(t)| < \varepsilon. \qquad (4.4)$$

Setting $\|x\| = \sup_t |x(t)|$ and defining Id to be an identity mapping, we may write this in a bit more compact form as

$$d(x, y) = \inf_{v \in \Upsilon} \{\max(\|v - Id\|, \|x - y \circ v\|)\} \qquad (4.5)$$

Note that the above indeed defines a metric: It may be shown that each cadlag function on $[0,1]$ is bounded thus $d(x, y)$ is always finite (since we may take $v(t) = t$). Of course $d(x, y) \geq 0$ and $d(x, y) = 0$ implies that for each t either $x(t) = y(t)$ or $x(t) = y(t^-)$ which in turn implies that $x = y$. Note that if $v \in \Upsilon$ then also $v^{-1} \in \Upsilon$ as well as if $v_1, v_2 \in \Upsilon$ then also $v_1 \circ v_2 \in \Upsilon$. The fact that $d(x, y) = d(y, x)$ follows directly from (4.5). Finally, the triangle inequality follows since

$$\|v_1 \circ v_2 - Id\| \leq \|v_1 - Id\| + \|v_2 - Id\|$$

as well as

$$\|x - z \circ v_1 \circ v_2\| \leq \|x - y \circ v_2\| + \|y - z \circ v_1\|.$$

We may now topologize the collection of all cadlag functions on $[0,1]$ with the metric d.

Definition 4.3.1 (Skorohod space). *The space of all cadlag functions on $[0, 1]$ with values in $\mathbf{R}^k$ with the topology induced by the metric d given by (4.5) is called Skorohod space and denoted by $D_{\mathbf{R}^k}$.*

From the above definition it follows that the sequence of elements x_n converges to x in the sense of the (topology of) space $D_{\mathbf{R}^k}$ if there exist functions $v_n \in \Upsilon$ such that $\lim_n x_n(v_n(t)) = x(t)$ uniformly in t and $\lim v_n(t) = t$ uniformly in t. Thus if $d_c(x_n, x) \to 0$ where d_c is given by (4.3), then x_n converges to x in $D_{\mathbf{R}^k}$, that is, $d(x_n, x) \to 0$. Conversely, since

$$|x_n(t) - x(t)| \leq |x_n(t) - x(v_n(t))| + |x(v_n(t)) - x(t)|$$

therefore $d(x_n, x) \to 0$ implies $d_c(x_n, x) \to 0$. Thus for continuous x_n, x the two metrics are equivalent. However, in general they are not as the example below illustrates.

Example 4.3.1 (Convergence in $D_{\mathbf{R}}$). Let $\alpha \in [0, 1)$ and take $x_n = I_{[0,\alpha+1/n)}$. Then the sequence (x_n) converges in (the Skorohod topology of) $D_{\mathbf{R}}$, namely $x_n = I_{[0,\alpha+1/n)} \to x = I_{[0,\alpha)}$ but on the other hand $x_n(t) \not\to x(t)$ for $t = \alpha$.

The important property of the Skorohod space $D_{\mathbf{R}^k}$ is that it is a separable metric space. The proof of this fact may be found e.g., in Billingsley (1999, p. 128).

In order to relate the convergence in the space $D_{\mathbf{R}^k}$ with our results on weak convergence for random permanents discussed in the previous chapter, we need to relate the convergence of the random elements $x(t)$ to the convergence of their finite dimensional projections $\Delta_{t_1 \ldots, t_k}(x) = x((t_1), \ldots, x(t_k))$. We note that since every function in Υ fixes the points $0, 1$, thus the special projections $\Delta_t(x)$ are always continuous for $t = 0, 1$ and hence in particular we have the following lemma.

Lemma 4.3.1. *The weak convergence of the random elements $x_n \xrightarrow{d} x$ in $D_{\mathbf{R}^k}$ implies convergence in distribution of $\mathbf{R}^k$-valued random vectors*

$$x_n(1) \xrightarrow{d} x(1).$$

Let us illustrate the usefulness of the above statement in deriving classical limit theorems. Consider the following functional version of the central limit theorem known as the Donsker invariance principle.

Theorem 4.3.1 (Donsker's theorem). *Let $Y_{l,k}$, $k = 1, \ldots, l$, and $l = 1, 2, \ldots$ be a double array of square integrable, zero-mean, row-wise independent real random variables. Assume that*

$$\lim_{l \to \infty} \frac{1}{l} \sum_{k=1}^{l} Var\, Y_{l,k} = \sigma^2, \tag{4.6}$$

$$\forall \varepsilon > 0 \quad \lim_{l \to \infty} \frac{1}{l} \sum_{k=1}^{l} E\, Y_{l,k}^2 I\{|Y_{l,k}| > \sigma \varepsilon \sqrt{l}\} = 0. \tag{4.7}$$

For $t \in [0, 1]$ define

$$Y_t^{(l)} = \frac{1}{\sigma \sqrt{l}} \sum_{k=1}^{[lt]} Y_{l,k}.$$

Then the weak convergence

$$Y^{(l)} \xrightarrow{d} \mathbf{B} \qquad l \to \infty$$

holds in the sense of the Skorohod space $D_{\mathbf{R}}([0, 1])$ where $\mathbf{B} = (B_t)_{t \in [0,1]}$ is the standard Brownian motion.

The theorem is a special case of the general result on the invariance principle for the triangular arrays of cadlag martingales. For the proof, see e.g., Billingsley (1999, p. 194). In view of Lemma 4.3.1 the above theorem gives us the following corollary.

Corollary 4.3.1 (Central limit theorem for triangular arrays). *Let* $Y_{l,k}$, $k = 1, \ldots, l$, *and* $l = 1, 2, \ldots$ *be a double array of square integrable, zero-mean, row-wise independent real random variables. Assume that* (4.6) *and* (4.7) *hold true. Then*

$$\frac{1}{\sigma \sqrt{l}} \sum_{k=1}^{l} Y_{l,k} \overset{d}{\to} \mathcal{N} \qquad l \to \infty.$$

Note that (4.7) is simply the Lindeberg condition (LC) of Theorem 3.3.1.

4.4 Permanent Stochastic Process

Recall that under the assumptions (A1-A2) introduced in Chapter 1 we are concerned with the setting where $\mathbb{X}^{(m,n)} = [X_{i,j}]$ is an $m \times n$ $(m \leq n)$ real random matrix of square integrable components and such that its columns are build from the first m terms of independent identically distributed (iid) sequences $(X_{i,1})_{i \geq 1}$, $(X_{i,2})_{i \geq 1}$, $\ldots$, $(X_{i,n})_{i \geq 1}$ of exchangeable random variables. As already noted, under these assumptions all entries of the matrix $\mathbb{X}^{(m,n)}$ are identically distributed, albeit not necessarily independent. Recall that we denote for $i, k = 1, \ldots, m$ and $j = 1, \ldots, n$ $\sigma^2 = Var\, X_{i,j}$ and $\rho = Corr(X_{k,j}, X_{i,j})$ where $\rho \geq 0$. As in the previous chapter we assume that $\mu \neq 0$ and denote additionally by $\gamma = \sigma/|\mu|$ the variation coefficient. In what follows $m = m_n \leq n$ is always a non-decreasing sequence.

In the setting described we proceed as follows. First, we extend the definition of a permanent function to a stochastic process (in the sequel referred to as "permanent stochastic process" or PSP) in a manner similar to that of extending the concept of a U-statistic defined in Chapter 1 to a "U-process" (see, Lee 1990). The fact that such a construction is possible is a consequence of the orthogonal decomposition (3.13). Second, in order to obtain the functional limit theorem for permanents, we extend some of the ideas of Chapter 3 to derive a general limit theorem for an elementary symmetric polynomial process (ESPP) based on a triangular array of row-wise independent, zero mean, square integrable random variables. To this end we will adopt from Kurtz and Protter (1991) an idea of representing the ESPP as a multiple stochastic integral.

Consider again the decomposition of a permanent given by (3.13). Extending the definition of orthogonal components $U_c^{(m,n)}$ $(c = 1,2,\ldots)$ to random functions is straightforward. For any real number x let $[x]$ denote the largest integer less or equal to x. Then for any c = 1,2,... and for any $t \in [0, 1]$ such that $c \leq [nt]$ let us define

$$U_c^{(m,n)}(t) = \frac{1}{\binom{n}{c}\binom{m}{c}c!} \sum_{1 \leq i_1 < \ldots < i_c \leq m} \sum_{1 \leq j_1 < \ldots < j_c \leq [nt]} Per\,[\tilde{X}_{i_u,j_v}]_{\substack{u = 1,\ldots,c \\ v = 1,\ldots,c}}$$

for $[\tilde{X}_{i,j}] = [X_{i,j}/\mu - 1]$ and put $U_c^{(m,n)}(t) = 0$ if $c > [nt]$.

Now, the decomposition (3.13) suggests that on the interval $[0,1]$ we may define the stochastic process associated with $\text{Per}\,(\mathbb{X}^{(m,n)})$, denoted by $\left(Per\,\mathbb{X}^{(m,n)}(t)\right)_{t\in[0,1]}$, as follows.

Definition 4.4.1 (Permanent stochastic process). *The permanent stochastic process is defined by*

$$Per\,\mathbb{X}^{(m,n)}(t) = \binom{n}{m} m!\,\mu^m \left[1 + \sum_{c=1}^{m}\binom{m}{c} U_c^{(m,n)}(t)\right] \qquad t \in [0,1].$$

It follows from (3.13) that the trajectory of $(Per\,\mathbb{X}^{(m,n)}(t))_{t\in[0,1]}$ coincides with a permanent function for $t = 1$, i.e.

$$Per\,\mathbb{X}^{(m,n)}(1) = \text{Per}\,(\mathbb{X}^{(m,n)})\,. \tag{4.8}$$

4.5 Weak Convergence of Stochastic Integrals and Symmetric Polynomials Processes

Naturally, to define an elementary symmetric polynomial process (ESPP) $(S^{(j,l)})$ based on any triangular array $[Y_{l,k}]$ of random variables, we refer to the notion of an elementary symmetric polynomial discussed already in Section 1.4 of Chapter 1 and Section 3.2.3 of Chapter 3. For any $j \leq l$ and for any $t \in [0,1]$ we define the process $S^{(j,l)}$ as

$$S_t^{(j,l)} = \begin{cases} S_{[lt]}(j), & \text{for } \frac{j}{l} \leq t \leq 1, \\ 0, & \text{for } 0 < t < \frac{j}{l} \end{cases}.$$

Recall that

$$S_{[lt]}(j) = \sum_{1\leq k_1 < ... < k_j \leq [lt]} Y_{l,k_1} \ldots Y_{l,k_j},$$

is an elementary symmetric polynomial of degree j based on $[lt]$ variables.

In this section we derive a general result on weak convergence of ESPPs based on any row-wise independent triangular array of zero mean, square integrable random variables. The result can be viewed as the extension of Donsker's invaraince principle (Theorem 4.3.1).

Theorem 4.5.1. *Let $Y_{l,k}$, $k = 1,\ldots,l$, $l = 1,2,\ldots$, be a double array of square integrable, zero-mean, row-wise independent random variables. Assume that (4.6) and (4.7) hold.*

Then, for any positive integer c,

$$\left[\left(\frac{S^{(1,l)}}{\sqrt{l}}\right),\ldots,\left(\frac{S^{(c,l)}}{(\sqrt{l})^c}\right)\right] \xrightarrow{d} \tag{4.9}$$

$$\xrightarrow{d} \left[\left(\frac{\sigma t^{1/2}}{1!} H_1 \left(\frac{B_t}{\sqrt{t}} \right) \right)_{t \in [0,1]}, \dots, \left(\frac{\sigma^c t^{c/2}}{c!} H_c \left(\frac{B_t}{\sqrt{t}} \right) \right)_{t \in [0,1]} \right]$$

as $l \to \infty$ in the Skorohod space $D_{\mathbf{R}^c}([0,1])$, where H_j is the j-th Hermité polynomial, $j = 1, 2, \dots$, and $\mathbf{B} = (B_t)_{t \in [0,1]}$ is the standard Brownian motion.

The asymptotic distribution of $l^{-c/2} S^{(l)}_{[lt]}(c)$ for fixed c and iid random variables appears for the first time in Móri and Székely (1982). The relation (4.9), again for iid random variables only, appears in Kurtz and Protter (1991) as one of the examples of applicability of their general result.

Note that the above theorem can be viewed as a functional version of the result on elementary symmetric polynomial given in Theorem 3.3.1. Several different ways of proving the above statement (4.9) are possible. Here we choose to apply an argument based on a quite general result on the convergence of stochastic integrals, due to Kurtz and Protter (1991). We have chosen this approach below since the application of the stochastic integral theory allows us to present a more concise argument, avoiding many technical details that needed to be addressed in our discussions in Chapter 3.

In order to proceed, we have to recall some basic facts and definitions. Let $\mathbb{F} = (\mathcal{F}_t)_{t \in [0,1]}$ be a filtration in a probability space $(\Omega, \mathcal{F}, P)$. If $Y = (Y_t)_{t \in [0,1]}$ is a right continuous $\mathbb{F}$ martingale then its quadratic variation process $([Y]_t)_{t \in [0,1]}$ is defined for any $t \in [0,1]$ as the limit

$$\sum_{k=1}^{n} (Y_{u_{k+1}^{(n)}} - Y_{u_k^{(n)}})^2 \xrightarrow{P} [Y]_t,$$

where $(u_k^{(n)})$ is a sequence of non-random partitions of $[0, t]$ such that $\max_k (u_{k+1}^{(n)} - u_k^{(n)}) \to 0$ as $n \to \infty$.

If Y is an $\mathbb{F}$ martingale with trajectories in $D_{\mathbf{R}}$ and $X = (X_t)_{t \in [0,1]}$ is an $\mathbb{F}$ adapted process with trajectories in $D_{\mathbf{R}^k}$ then we define *the stochastic integral* $(\int_0^t X_s \, dY_s)_{t \in [0,1]}$ as a limit in probability of the Riemann-Stieltjes sums

$$\sum X_{t_i(n)} (Y_{t_{i+1}^{(n)}} - Y_{t_i(n)}) \xrightarrow{P} \int_0^t X_s \, dY_s,$$

for non-random partitions $(t_i^{(n)})$ of $[0, t]$ such that $\max(t_{i+1}^{(n)} - t_i^{(n)})$ converges to zero. The integral exists if the respective limit in probability of the Riemann-Stieltjes sums does not depend on the choice of the sequence $(t_i^{(n)})$. We will use the notation $\int X \, dY$ for the process $\left(\int_0^t X_s \, dY_s \right)_{t \in [0,1]}$.

The version of a convergence result for stochastic integrals which suits our purposes can be formulated in the following way.

Theorem 4.5.2. *Assume that* (X, Y), (X_1, Y_1), (X_2, Y_2), $\dots$ *are pairs of stochastic processes such that trajectories of the first components are in* $D_{\mathbf{R}^k}$

and trajectories of the second components are in $D_{\mathbf{R}}$. *Let* Y_n *be a square integrable martingale with respect to a filtration* $\mathbb{F}_n$ *and* $\sup_{n \geq 1} E([Y_n]_t) < \infty$ *for
any* $t \in [0, 1]$, *where* $([Y_n]_t)_{t \in [0,1]}$ *is the quadratic variation process for* Y_n,
$n = 1, 2, \ldots$. *Assume also that the process* X_n *is* $\mathbb{F}_n$ *adapted,* $n = 1, 2, \ldots$ *If*
(X_n, Y_n) *converges weakly to* (X, Y) *in* $D_{\mathbf{R}^{k+1}}$ *then* $(X_n, Y_n, \int X_n \, dY_n)$ *converges weakly to* $(X, Y, \int X \, dY)$ *in* $D_{\mathbf{R}^{2k+1}}$.   □

The proof of this theorem may be found in Kurtz and Protter (1991).

We shall use Theorem 4.5.2 in order to give a concise proof of Theorem 4.5.1. It is based on the idea of representing elementary symmetric polynomials via iterative stochastic integrals.

Proof (of Theorem 4.5.1). Define for any $k = 1, 2, \ldots, c$

$$Y_n^{(k)} = \frac{S^{(k,n)}}{(\sqrt{n})^k}.$$

Observe that

$$Y_n^{(k)} = \int Y_n^{(k-1)} \, dY_n^{(1)}.$$

Also define

$$Y^{(k)} = \left(Y_t^{(k)} \right)_{t \in [0,1]} = \left(\frac{\sigma^k t^{k/2}}{k!} H_k \left(\frac{B_t}{\sqrt{t}} \right) \right)_{t \in [0,1]}.$$

It is well known that

$$Y^{(k)} = \int Y^{(k-1)} \, dY^{(1)}.$$

Let $Z_n^{(k)} = [Y_n^{(1)}, \ldots, Y_n^{(k)}]$, $n = 1, 2, \ldots$ and let $Z^{(k)} = [Y^{(1)}, \ldots, Y^{(k)}]$, $k = 1, 2, \ldots, c$.
    To prove that

$$Z_n^{(k)} \xrightarrow{d} Z^{(k)},$$

in the Skorohod space $D_{\mathbf{R}^k}$, $k = 1, \ldots, c$, we use induction with respect to k.
    The result for $k = 1$ follows in view of the conditions (4.6) and (4.7)
which imply, by the Donsker theorem for triangular arrays (Theorem 4.3.1),
that $Z_n^{(1)} = Y_n^{(1)} \xrightarrow{d} \mathbf{B} = Y^{(1)}$ in $D_{\mathbf{R}}$. Observe also that the process $Y_n^{(1)}$
is a martingale since it is based on the summation of independent random
variables. Moreover, since for a given n the process $Y_n^{(1)}$ just cumulates jumps
while remaining constant, in-between jumps, then its quadratic variation is

$$[Y_n^{(1)}]_t = n^{-1} \sum_{k=1}^{[nt]} Y_{n,k}^2.$$

Consequently,

$$\sup_{n} E([Y_n^{(1)}]_t) = \sup_{n} \frac{[nt]\sigma^2}{n} \leq t\sigma^2.$$

Let us also note that all the processes $Y_n^{(k)}$, $k = 1, \ldots, c$, are adapted to the same filtration, generated by $Y_n^{(1)}$, $n = 1, 2, \ldots$, and similarly, the processes $Y^{(k)}$, $k = 1, \ldots, c$, are adapted to the same filtration, generated by the Wiener process $\mathbf{B}$.

Assume now that $Z_n^{(k-1)} \xrightarrow{d} Z^{(k-1)}$ in $D_{\mathbf{R}^{k-1}}$. Then, obviously,

$$\left(Z_n^{(k-1)}, Y_n^{(1)} \right) \xrightarrow{d} \left(Z^{(k-1)}, Y^{(1)} \right)$$

in $D_{\mathbf{R}^k}$ since the limiting process is continuous a.s. Consequently, by Theorem 4.5.2 it follows that

$$\left(Z_n^{(k-1)}, Y_n^{(1)}, \int Z_n^{(k-1)} \, dY_n^{(1)} \right) \xrightarrow{d} \left(Z^{(k-1)}, Y^{(1)}, \int Z^{(k-1)} \, dY^{(1)} \right).$$

But the above convergence yields immediately the final result since

$$\left(Y_n^{(1)}, \int Z_n^{(k-1)} \, dY_n^{(1)} \right) = Z_n^{(k)}$$

and $(Y^{(1)}, \int Z^{(k-1)} \, dY^{(1)}) = Z^{(k)}$. $\qquad\square$

4.6 Convergence of the Component Processes

Recall that by the definition of PSP (Definition 4.4.1)

$$\frac{Per \, \mathbb{X}^{(m,n)}(t)}{\binom{n}{m} m! \mu^m} = 1 + \sum_{c=1}^{m} \binom{m}{c} U_c^{(m,n)}(t) \qquad t \in [0, 1].$$

We shall first establish the limiting result for processes $U_c^{(m,n)}$ ($c = 1, 2, \ldots$).

To this end, let us consider for any $c = 1, 2, \ldots$ a process $(W_c^{(n)}(t))$ defined by the following rescaling of $U_c^{(m,n)}(t)$.

$$W_c^{(n)}(t) = \binom{n}{c}\binom{m}{c} c! \, U_c^{(m,n)}(t) = \sum_{1 \leq i_1 < \ldots i_c \leq m} \sum_{1 \leq j_1 < \ldots j_c \leq [nt]} Per \left[\tilde{X}_{i_u, j_v} \right]_{\substack{u = 1, \ldots, c \\ v = 1, \ldots, c}}$$

for $t \in [c/n, 1]$ and $W_c^{(n)}(t) = 0$ for $t \in [0, c/n)$. The reminder of this section is devoted to the proof of the following functional analogue of Theorem 3.3.1 from Chapter 3.

Theorem 4.6.1. *Let c be an arbitrary positive integer. Assume that $m = m_n \to \infty$ as $n \to \infty$.*

If $\rho = 0$ then

$$\left[\frac{W_1^{(n)}(t)}{\sqrt{mn}}, \ldots, \frac{W_c^{(n)}(t)}{(\sqrt{mn})^c}\right]_{t \in [0,1]} \xrightarrow{d} \left[\frac{\gamma\sqrt{t}H_1\left(\frac{B_t}{\sqrt{t}}\right)}{1!}, \ldots, \frac{(\gamma\sqrt{t})^c H_c\left(\frac{B_t}{\sqrt{t}}\right)}{c!}\right]_{t \in [0,1]}$$

in $D_{\mathbf{R}^c}$.

If $\rho > 0$ then

$$\left[\frac{W_1^{(n)}(t)}{m\sqrt{n}}, \ldots, \frac{W_c^{(n)}(t)}{(m\sqrt{n})^c}\right]_{t \in [0,1]} \xrightarrow{d} \left[\frac{\gamma\sqrt{\rho t}H_1\left(\frac{B_t}{\sqrt{t}}\right)}{1!}, \ldots, \frac{(\gamma\sqrt{\rho t})^c H_c\left(\frac{B_t}{\sqrt{t}}\right)}{c!}\right]_{t \in [0,1]}$$

in $D_{\mathbf{R}^c}$.

Proof. Consider first the case $\rho = 0$. For an arbitrary fixed positive integer c define a process $V_c^{(n)}$ by

$$V_c^{(n)}(t) = n^{-c/2} \sum_{1 \leq j_1 < \ldots < j_c \leq [nt]} Y_{n,j_1} \ldots Y_{n,j_c}$$

for $t \in [c/n, 1]$, where

$$Y_{n,j} = m^{-1/2} \sum_{i=1}^{m} \tilde{X}_{i,j}, \quad j = 1, \ldots, n,$$

and $V_c^{(n)}(t) = 0$ for $t \in [0, c/n)$.

Observe that by Theorem 4.5.1 the convergence

$$[V_1^{(n)}(t), \ldots, V_c^{(n)}(t)]_{t \in [0,1]} \xrightarrow{d} \left[\frac{\gamma\sqrt{t}H_1\left(\frac{B_t}{\sqrt{t}}\right)}{1!}, \ldots, \frac{(\gamma\sqrt{t})^c H_c\left(\frac{B_t}{\sqrt{t}}\right)}{c!}\right]_{t \in [0,1]} \tag{4.10}$$

in $D_{\mathbf{R}^c}$ follows as soon as we verify (4.6) and the Lindeberg condition (4.7), which in our current setting takes the form

$$E(Y_{n,1}^2 I(|Y_{n,1}| > \sqrt{n}\varepsilon)) \to 0$$

for any $\varepsilon > 0$ as $n \to \infty$. This follows along the same lines like in the respective part of the proof of Theorem 3.3.1. Therefore (4.6) and (4.7) are satisfied and thus, via Theorem 4.5.1, also (4.10) holds true.

Observe that for any $k = 1, \ldots, c$

$$\frac{W_k^{(n)}(t)}{(\sqrt{mn})^k} = V_k^{(n)}(t) - \frac{R_k^{(n)}(t)}{(\sqrt{mn})^k}, \tag{4.11}$$

where $(R_k^{(n)}(t))$ is a process such that for any $t \in [k/n, 1]$ it is a sum of different products $\tilde{X}_{i_1,j_1} \ldots \tilde{X}_{i_k,j_k}$ where $1 \leq j_1 < \ldots < j_k \leq [nt]$, $(i_1, \ldots, i_k) \in \{1, \ldots, m\}^k$ and at least one of the indices $i_1, \ldots, i_k$ in the sequence $(i_1, \ldots, i_k)$ repeats. For $t \in [0, k/n)$ we define $R_k^{(n)}(t)$ to be equal to zero.

Note that $(R_k^{(n)}(t))_{t\in[0,1]}$ is a martingale for any fixed n, in view of the assumed independence of columns of $\mathbb{X}^{(m,n)}$. Consequently, by the maximal inequality (Theorem 2.1.1),

$$P\left(\sup_{t\in[0,1]} \frac{|R_k^{(n)}(t)|}{(\sqrt{mn})^k} > \varepsilon \right) \leq \frac{E\,(R_k^{(n)}(1))^2}{\varepsilon^2(mn)^k} = \frac{Var\,(R_k^{(n)}(1))}{\varepsilon^2(mn)^k}.$$

Now, repeating the argument used in the first part of the proof of Theorem 3.3.1 we arrive at

$$Var\left(\frac{R_k^{(n)}(1)}{(\sqrt{nm})^k} \right) \to 0,$$

as $n \to \infty$.

Consequently, the c-variate process

$$\left[\frac{R_1^{(n)}(t)}{\sqrt{mn}}, \ldots, \frac{R_c^{(n)}(t)}{(\sqrt{mn})^c} \right]^T_{t\in[0,1]} \xrightarrow{P} 0$$

in $D_{\mathbf{R}^c}$ and thus the first result follows immediately by (4.10) and (4.11). Thus the case $\rho = 0$ is completed.

Next, let us consider the case $\rho > 0$. As above, let us define for any $k = 1, 2, \ldots$ the process $(V_k^{(n)}(t))_{t\in[0,1]}$ by

$$V_k^{(n)}(t) = n^{-c/2} \sum_{1\leq j_1<\ldots<j_c\leq[nt]} Y_{n,j_1} \ldots Y_{n,j_c},$$

where

$$Y_{n,j} = \frac{1}{m} \sum_{i=1}^m \tilde{X}_{ij},$$

$j = 1, 2, \ldots, n$.

Similarly to the argument used in the second part of the proof of Theorem 3.3.1 we get

$$[V_1^{(n)}(t), \ldots, V_c^{(n)}(t)]_{t\in[0,1]} \xrightarrow{d} \left[\frac{(\sqrt{\rho t}\gamma)^1}{1!} H_1(B_t/\sqrt{t}), \ldots, \right.$$
$$\left. \frac{(\sqrt{\rho t}\gamma)^c}{c!} H_c(B_t/\sqrt{t}) \right]_{t\in[0,1]}. \quad (4.12)$$

Also as in the previous case of uncorrelated components, for any $k = 1, \ldots, c$, we have

$$\frac{W_k^{(n)}(t)}{m^k(\sqrt{n})^k} = V_k^{(n)}(t) - \frac{R_k^{(n)}(t)}{m^k(\sqrt{n})^k}. \tag{4.13}$$

Again we refer to the second part of the proof of Theorem 3.3.1 to conclude that

$$Var\left(\frac{R_k^{(n)}(1)}{m^k(\sqrt{n})^k}\right) \to 0$$

as $n \to \infty$. As before the process $(R_k^{(n)}(t))_{t \in [0,1]}$ is a martingale for fixed n and thus once again the maximal inequality yields

$$\left[\frac{R_1^{(n)}(t)}{m\sqrt{n}}, \ldots, \frac{R_c^{(n)}(t)}{(m\sqrt{n})^c}\right]_{t \in [0,1]}^T \xrightarrow{P} 0$$

in $D_{\mathbf{R}}$.

Now, the second assertion of the theorem follows from (4.12) and (4.13) and the proof of Theorem 4.6.1 is complete. $\qquad \square$

4.7 Functional Limit Theorems

Having established the result of Theorem 4.6.1 in the previous subsection, we are finally in a position to state and prove the two main theorems for a permanent stochastic process defined in Definition 4.4.1. As stated earlier, the present limit theorems could be viewed as the functional generalizations of the Theorems 3.4.2 and 3.4.3 from Chapter 3. In particular, due to (4.8) and the result of Lemma 4.3.1 they entail these results.

To begin, we consider the case when the entries of matrix $\mathbb{X}^{(\infty)}$ are uncorrelated.

Theorem 4.7.1. *Assume that $\rho = 0$ and let $(B_t)_{t \in [0,1]}$ denote the standard Brownian motion.*

If $m/n \to \lambda > 0$ as $n \to \infty$ then

$$\frac{(Per\,\mathbb{X}^{(m,n)}(t))_{t \in [0,1]}}{\binom{n}{m}m!\mu^m} \xrightarrow{d} \left(\exp\left(\sqrt{\lambda}\gamma B_t - \lambda t \gamma^2\right)\right)_{t \in [0,1]} \quad in\ D_{\mathbf{R}}. \tag{4.14}$$

If $m/n \to 0$ and $m = m_n \to \infty$ as $n \to \infty$ then

$$\sqrt{\frac{n}{m}}\left(\frac{(Per\,\mathbb{X}^{(m,n)}(t))_{t \in [0,1]}}{\binom{n}{m}m!\mu^m} - 1\right) \xrightarrow{d} \gamma(B_t)_{t \in [0,1]} \quad in\ D_{\mathbf{R}}. \tag{4.15}$$

Proof. Consider first the case $\lambda > 0$. For any n and any N such that $N < m_n$ define a process $\mathbf{S}_N^{(n)} = (S_N^{(n)}(t))_{t \in [0,1]}$ by

$$S_N^{(n)}(t) = 1 + \sum_{c=1}^{N} \binom{m}{c} U_c^{(m,n)}(t) = 1 + \sum_{c=1}^{N} \frac{(\sqrt{mn})^c}{\binom{n}{c} c!} \frac{W_c^{(n)}(t)}{(\sqrt{mn})^c}.$$

Observe that by the first assertion of Theorem 4.6.1 it follows that

$$\mathbf{S}_N^{(n)} \xrightarrow{d} \mathbf{G}_N = \left(\sum_{c=0}^{N} \frac{(\lambda t \gamma^2)^{c/2}}{c!} H_c \left(\frac{B_t}{\sqrt{t}} \right) \right)_{t \in [0,1]} \tag{4.16}$$

in $D_{\mathbf{R}}$ as $n \to \infty$, since for any $c = 1, 2, \dots$

$$\frac{(\sqrt{mn})^c}{\binom{n}{c} c!} \to \lambda^{c/2}.$$

Define also a process $\mathbf{T}_N^{(n)} = (T_N^{(n)}(t))_{t \in [0,1]}$ by

$$T_N^{(n)}(t) = \sum_{c=N+1}^{m_n} \binom{m}{c} U_c^{(m,n)}(t).$$

And let $\mathbf{G}_\infty = (G_\infty(t))_{t \in [0,1]}$ be defined by

$$(G_\infty(t))_{t \in [0,1]} = \left(\sum_{c=0}^{\infty} \frac{(\lambda t \gamma^2)^{c/2}}{c!} H_c \left(\frac{B_t}{\sqrt{t}} \right) \right)_{t \in [0,1]}$$
$$= \left(\exp \left(\sqrt{\lambda} \gamma B_t - \lambda t \gamma^2 \right) \right)_{t \in [0,1]}.$$

Let ϱ denote the Prokhorov distance between the random elements in $D_{\mathbf{R}}$. Then

$$\varrho(\mathbf{Z}^{(n)}, \mathbf{G}_\infty) \le \varrho(\mathbf{Z}^{(n)}, \mathbf{S}_N^{(n)}) + \varrho(\mathbf{S}_N^{(n)}, \mathbf{G}_N) + \varrho(\mathbf{G}_N, \mathbf{G}_\infty), \tag{4.17}$$

where $\mathbf{Z}^{(n)} = \mathbf{S}_N^{(n)} + \mathbf{T}_N^{(n)}$. Let us note that for any $t \in [0,1]$

$$Per\, \mathbb{X}^{(m,n)}(t) = \binom{n}{m} m!\, \mu^m\, Z^{(n)}(t)$$

and that (4.16) implies $\varrho(\mathbf{S}_N^{(n)}, \mathbf{G}_N) \to 0$ as $n \to \infty$ for any fixed N, as well as, that we have $\varrho(\mathbf{G}_N, \mathbf{G}_\infty) \to 0$ as $N \to \infty$. Therefore, in order to argue that $\varrho(\mathbf{Z}^{(n)}, \mathbf{G}_\infty) \to 0$ as $n \to \infty$ we only need to show that $\varrho(\mathbf{Z}^{(n)}, \mathbf{S}_N^{(n)})$ tends to zero uniformly in n as $N \to \infty$.

To this end observe that by the bound on the Prohorov distance given in Theorem 4.2.3 as well as the fact that the distance in the uniform metric d_C given by (4.3) bounds the Skorohod distance d, we have

$$\varrho(\mathbf{Z}^{(n)}, \mathbf{S}_N^{(n)}) \le \inf\{\varepsilon > 0 : P(\sup_{t\in[0,1]} |T_N^{(n)}(t)| > \varepsilon) \le \varepsilon\}.$$

On the other hand, as before, it is easy to see that due to the assumed independence of columns of $\mathbb{X}^{(m,n)}$ the process $\mathbf{T}_N^{(n)}$ is a martingale for fixed n, N and thus, via the maximal inequality and the relations (3.14) and (3.15) for $\rho = 0$, we have for any $\varepsilon > 0$

$$P(\sup_{t\in[0,1]} |T_N^{(n)}(t)| > \varepsilon) \le \varepsilon^{-2} Var\, T_N^{(n)}(1) = \varepsilon^{-2} \sum_{c=N+1}^{m} \frac{\binom{m}{c}\gamma^2}{\binom{n}{c}c!}$$

$$\le \varepsilon^{-2} \sum_{c=N+1}^{\infty} \frac{\gamma^{2c}}{c!} = \varepsilon^{-2}\, \alpha_N \to 0$$

as $N \to \infty$, which entails

$$\varrho(\mathbf{Z}^{(n)}, \mathbf{S}_N^{(n)}) \le \sqrt[3]{\alpha_N} \to 0,$$

uniformly in n, as $N \to \infty$ and hence, by (4.17) that $\varrho(\mathbf{Z}^{(n)}, \mathbf{G}_\infty) \to 0$ as $n \to \infty$.

Thus, we conclude that $\mathbf{Z}^{(n)}$ converges weakly to $\mathbf{G}_\infty$ in $D_{\mathbf{R}}$.

For the proof in the case $\lambda = 0$, let us observe that

$$\sqrt{\frac{n}{m}} \sum_{c=1}^{m} \binom{m}{c} U_c^{(m,n)}(t) = \frac{W_1^{(n)}(t)}{\sqrt{mn}} + R^{(m,n)}(t),$$

where

$$R^{(m,n)}(t) = \sqrt{\frac{n}{m}} \sum_{c=2}^{m} \binom{m}{c} U_c^{(m,n)}(t).$$

Observe that due to the independence of columns of the matrix $\mathbb{X}$ the process $(R^{(m,n)}(t))$ is a martingale for fixed values of m, n. Consequently, by the maximal inequality we get

$$P(\sup_{t\in[0,1]} |R^{(m,n)}(t)| > \varepsilon) \le \varepsilon^{-2} Var\, R^{(m,n)}(1) = \varepsilon^{-2} \frac{n}{m} \sum_{c=2}^{m} \frac{\binom{m}{c}\gamma^{2c}}{\binom{n}{c}c!}$$

$$\le \frac{m}{n} \exp(\gamma^2) \to 0$$

since $m/n \to \infty$ as $n \to \infty$ and conclude that $(R^{(m,n)}(t))$ converges to zero in probability in $D_{\mathbf{R}}$. Hence, the final result follows now directly by Theorem 4.6.1. $\square$

A quick inspection of the proof of the second part of the theorem reveals that a slight modification of the argument shows validity of the second part of the hypothesis also for m remaining fixed. In particular, let us note that for $m = 1$ we obtain simply Donsker's theorem (Theorem 4.3.1).

The result of Theorem 4.7.1 is complemented with the corresponding one in the case when $\rho > 0$.

Theorem 4.7.2. *Assume that $\rho > 0$ and, as before, let $(B_t)_{t\in[0,1]}$ denote the standard Brownian motion.*

If $m^2/n \to \lambda > 0$ as $n \to \infty$ then

$$\frac{(Per\, \mathbb{X}^{(m,n)}(t))_{t\in[0,1]}}{\binom{n}{m}m!\,\mu^m} \xrightarrow{d} \left(\exp\left(\sqrt{\lambda\rho}\,\gamma B_t - \lambda\rho t\gamma^2\right)\right)_{t\in[0,1]} \qquad in\ D_\mathbf{R}.$$

If $m^2/n \to 0$ and $m = m_n \to \infty$ as $n \to \infty$ then

$$\frac{\sqrt{n}}{m}\left(\frac{(Per\, X^{(m,n)}(t))_{t\in[0,1]}}{\binom{n}{m}m!\,\mu^m} - 1\right) \xrightarrow{d} \sqrt{\rho}\,\gamma(B_t)_{t\in[0,1]} \qquad in\ D_\mathbf{R}.$$

Proof. The proof of the result parallels, to large extent, that of Theorem 4.7.1. As before, consider first the case $\lambda > 0$. For any n and N such that $N < m_n$ define a process $\mathbf{S}_N^{(n)} = (S_N^{(n)}(t))_{t\in[0,1]}$ by

$$S_N^{(n)}(t) = 1 + \sum_{c=1}^{N}\binom{m}{n}U_c^{(m,n)}(t) = 1 + \sum_{c=1}^{N}\frac{(\sqrt{m}\,n)^c}{\binom{n}{c}c!}\frac{W_c^{(n)}(t)}{(\sqrt{m}\,n)^c}.$$

Observe that by the second assertion of Theorem 4.6.1 it follows that

$$\mathbf{S}_N^{(n)} \xrightarrow{d} \mathbf{G}_N = \left(\sum_{c=0}^{N}\frac{(\lambda\rho t\gamma^2)^{c/2}}{c!}H_c(B_t/\sqrt{t})\right)_{t\in[0,1]}$$

in $D_\mathbf{R}$ as $n \to \infty$, since for any $c = 1, 2, \ldots$

$$\frac{(\sqrt{n}\,m)^c}{\binom{n}{c}c!} \to \lambda^{c/2}.$$

Define also a process $\mathbf{T}_N^{(n)} = (T_N^{(n)}(t))_{t\in[0,1]}$ by

$$T_N^{(n)}(t) = \sum_{c=N+1}^{m_n}\binom{m}{c}U_c^{(m,n)}(t).$$

And let $\mathbf{G}_\infty = (G_\infty(t))_{t\in[0,1]}$ be defined by

$$(G_\infty(t))_{t\in[0,1]} = \left(\sum_{c=0}^{\infty}\frac{(\lambda\rho t\gamma^2)^{c/2}}{c!}H_c(B_t/t)\right)_{t\in[0,1]}$$

$$= \exp\left(\sqrt{\lambda\rho}\,\gamma B_t - \lambda\rho t\gamma^2\right)_{t\in[0,1]}.$$

As before, if ϱ denotes the Prokhorov distance between the random elements in $D_{\mathbf{R}}$, then

$$\varrho(\mathbf{Z}^{(n)}, \mathbf{G}_\infty) \leq \varrho(\mathbf{Z}^{(n)}, \mathbf{S}_N^{(n)}) + \varrho(\mathbf{S}_N^{(n)}, \mathbf{G}_N) + \varrho(\mathbf{G}_N, \mathbf{G}_\infty).$$

where $\mathbf{Z}^{(n)} = \mathbf{S}_N^{(n)} + \mathbf{T}_N^{(n)}$. In order to complete the argument along the lines of the first part of the proof of Theorem 4.7.1 we only need to argue that for any $\varepsilon > 0$

$$P(\sup_{t \in [0,1]} |T_N^{(n)}(t)| > \varepsilon) \to 0,$$

uniformly in n as $N \to \infty$. To this end note that again by the martingale property and the maximal inequality as well as by the relations (3.14) and (3.15)

$$P(\sup_{t \in [0,1]} |T_N^{(n)}(t)| > \varepsilon)$$

$$\leq \varepsilon^{-2} Var\, T_N^{(n)}(1) = \varepsilon^{-2} \frac{n}{m^2} \sum_{c=N+1}^{m} \frac{\binom{m}{c} \gamma^{2c}}{\binom{n}{c}} \sum_{r=0}^{c} \frac{1}{r!} \binom{m-r}{c-r} (1-\rho)^r \rho^{c-r}$$

$$\leq \varepsilon^{-2} \exp(1) \sum_{c=N+1}^{m} \left(\frac{m^2}{n}\right)^c \frac{\gamma^{2c}}{c!},$$

in view of the inequalities, by now familiar, $\binom{m-r}{c-r} \leq \binom{m}{c}$ and $c! \frac{\binom{m}{c}^2}{\binom{n}{c}} \leq \left(\frac{m^2}{n}\right)^c$ for $0 \leq r \leq c \leq m$. Therefore, for n large enough to have $m/n \leq 2\lambda$

$$P(\sup_{t \in [0,1]} |T_N^{(n)}(t)| > \varepsilon) \leq \varepsilon^{-2} \exp(1) \sum_{c=N+1}^{\infty} \frac{(2\lambda\gamma^2)^c}{c!} = \varepsilon^{-2}$$

Consequently, the result follows since $\alpha_N \to 0$ as $N \to \infty$.

The proof in the case $\lambda = 0$, follows similarly to the second part of the proof of Theorem 4.7.1 with obvious modifications. Consider

$$\frac{\sqrt{n}}{m} \sum_{c=1}^{m} \binom{m}{c} U_c^{(m,n)}(t) = \frac{W_1^{(n)}(t)}{m\sqrt{n}} + R^{(m,n)}(t),$$

where

$$R^{(m,n)}(t) = \frac{\sqrt{n}}{m} \sum_{c=2}^{m} \binom{m}{c} U_c^{(m,n)}(t).$$

The process $(R^{(m,n)}(t))$ is a martingale for fixed values of m, n and thus

$$P(\sup_{t \in [0,1]} |R^{(m,n)}(t)| > \varepsilon) \leq \varepsilon^{-2} Var\, R^{(m,n)}(1) \to 0,$$

follows, in view of

$$Var\, R_{m,n}(1) \leq \frac{m^2}{n} \exp(1 + \gamma^2)$$

if only n is large enough to have $m^2/n < 1$. To end the proof we refer again to the final part of the proof of Theorem 4.7.1. $\square$

As before, let us note that the above result remains valid for m being a constant (see the remark after the proof of Theorem 4.7.1).

4.8 Bibliographic Details

The general theory on weak convergence in metric spaces is discussed for instance in and excellent classical review book Billingsley (1999) (see also Billingsley 1995). In the context of Markov processes the theory of Skorohod spaces and Prohorov metric is also reviewed in Ethier and Kurtz (1986). The Donsker invariance principle is classical and can be found for instance in Billingsley (1999). For the results on weak convergence of stochastic integrals except of Kurtz and Protter (1991), one can consult also the important paper by Jakubowski et al. (1989). The paper by Kurtz and Protter (1991) is of special interest because except of general results on convergence of stochastic integrals it also contains some examples applicable to P-statistics theory and in particular a very special case of Theorem 4.5.1. The first result on convergence of ESPP belongs to Móri and Székely (1982). The results on weak convergence of elementary symmetric polynomial processes and of permanent processes presented in this chapter are taken from Rempała and Wesołowski (2005a). Readers are referred to that paper for further details and discussions.

5

Weak Convergence of P-statistics

In this chapter we shall prove a result on the weak convergence of generalized permanents which extends the results on random permanents presented in Chapter 3. Herein we return to the discussions of Chapter 1 where we introduced a matrix permanent function as a way to describe properties of perfect matchings in bipartite graphs. Consequently, the asymptotic properties which will be developed in this chapter for P-statistics can be immediately translated into the language of the graph theory as the properties of matchings in some bipartite random graphs. This will be done in Section 5.3, where we shall revisit some of the examples introduced in Chapter 1. In order to establish the main results of this chapter we will explore the path connecting the asymptotic behavior of U- and P-statistics. An important mathematical object which will be encountered here is a class of real random variables known as multiple Wiener-Itô integrals. The concept of the Wiener-Itô integral is related to that of a stochastic integral with respect to martingales introduced in Chapter 4, though its definition adopted in this chapter is somewhat different - it uses Hermité polynomial representations. It will be introduced in the next section. We shall start our discussion of asymptotics for P-statistics by first introducing the classical result for U-statistics with fixed kernel due to Dynkin and Mandelbaum, then obtaining a limit theorem for U-statistics with kernels of increasing order, and finally extending the latter to P-statistics.

The main theoretical results herein are given in Section 5.2 as Theorems 5.2.1 and 5.2.2 and describe the asymptotics for P-statistics of random matrices with independent identically distributed random entries as the number of columns and rows increases. The result of Theorem 5.2.2 is used in Section 5.3 in order to derive some asymptotic distributions for several functions of perfect matchings in random bipartite graphs under the assumption that the corresponding matrices of weights $\mathbb{X}$ have independent and identically distributed entries (e.g., the edges appear randomly or are randomly colored). Throughout this chapter we therefore strengthen the exchangeability assumptions on the entries of matrix $\mathbb{X}^{(\infty)}$ given by (A1)-(A2) in Chapter 1 to that of independence and identical distribution of the entries.

5.1 Multiple Wiener-Itô Integral as a Limit Law for U-statistics

5.1.1 Multiple Wiener-Itô Integral of a Symmetric Function

In this section we introduce the multiple Wiener-Itó integral, the object which appears as the limit in distribution for properly normalized P-statistics.

Let X be a random variable on a probability space $(\Omega, \mathcal{F}, P)$ having the distribution ν. Let $\mathcal{J} = \{J_1(\phi): \phi \in L^2(\mathbf{R}, \mathcal{B}, \nu)\}$ be a Gaussian system (possibly on a different probability space $(\widetilde{\Omega}, \widetilde{\mathcal{F}}, \mathcal{P})$ with the expectation operator denoted by $\mathcal{E}$) with zero means, unit variances and covariances $\mathcal{E}(J_1(\phi)J_1(\psi)) = E(\phi(X)\psi(X))$ for any $\phi, \psi \in L^2(\mathbf{R}, \mathcal{B}, \nu)$.

For any functions $f_i : \mathcal{U}_i \to V$, where $\mathcal{U}_i$ is an arbitrary set and V is an algebraic structure with multiplication, $i = 1, \ldots, m$, the tensor product $f_1 \otimes \ldots \otimes f_m : \mathcal{U}_1 \times \ldots \times \mathcal{U}_m \to V$ is defined as

$$f_1 \otimes \ldots \otimes f_m(\mathbf{x}) = f_1(x_1) \ldots f_m(x_m)$$

for any $\mathbf{x} = (x_1, \ldots, x_m) \in \mathcal{U}_1 \times \ldots \times \mathcal{U}_m$. In particular, for $f : \mathcal{U} \to V$ we have

$$f^{\otimes m}(\mathbf{x}) = f(x_1) \ldots f(x_m)$$

for any $\mathbf{x} = (x_1, \ldots, x_m) \in \mathcal{U}^m$.

For $\phi^{\otimes m}$, where $\phi \in L^2(\mathbf{R}, \mathcal{B}, \nu)$ is such that $E\phi(X) = 0$ and $E\phi^2(X) = 1$, the m-th multiple Wiener-Itô integral J_m is defined as

$$J_m(\phi^{\otimes m}) = \frac{1}{\sqrt{m!}} H_m\left(J_1(\phi)\right) ,$$

where H_m is the m-th monic Hermité polynomial (see Definition 3.2.2).

Moreover the linearity property is imposed on J_m, i.e. for any $\phi, \psi \in L^2(\mathbf{R}, \mathcal{B}, \nu)$, such that $E\,\phi(X) = 0 = E\,\psi(X)$ and $E\,\phi^2(X) = 1 = E\,\psi^2(X)$ and for any real numbers a and b it is required that

$$J_m(a\phi^{\otimes m} + b\psi^{\otimes m}) = aJ_m(\phi^{\otimes m}) + bJ_m(\psi^{\otimes m}).$$

Consequently, J_m is linear on the space

$$\mathcal{T}_s^{(m)} = span\left(\phi^{\otimes m} : \phi \in L^2(\mathbf{R}, \mathcal{B}, \nu), \ E\,\phi(X) = 0, \ E\,\phi^2(X) = 1\right)$$

which is a subspace of the space of symmetric functions $L_s^2(\mathbf{R}^m, \mathcal{B}_m, \nu_m) \subset L^2(\mathbf{R}^m, \mathcal{B}_m, \nu_m)$. Here $\mathcal{B}_m$ is the Borel σ-algebra in $\mathbf{R}^m$ and $\nu_m = \nu^{\otimes m}$ is the product measure.

As noted in Chapter 3 Section 3.2.3 it is well known that Hermité polynomials are orthogonal basis in the space $L^2(\mathbf{R}, \mathcal{B}, P_\mathcal{N})$, where $\mathcal{N}$ is a standard normal variable and $E(H_m^2(\mathcal{N})) = m!$, $m = 0, 1, \ldots$ Note that for any $\phi, \psi \in L^2(\mathbf{R}, \mathcal{B}, \nu)$, such that $E\,\phi(X) = 0 = E\,\psi(X)$ and $E\,\phi^2(X) = 1 = E\,\psi^2(X)$

$$m!\mathcal{E}\,J_m^2\left(a\phi^{\otimes m} + b\psi^{\otimes m}\right) = m!\mathcal{E}\left(aJ_m(\phi^{\otimes m}) + bJ_m(\psi^{\otimes m})\right)^2$$

$$= a^2\,\mathcal{E}\,H_m^2(J_1(\phi)) + 2ab\,\mathcal{E}\,H_m(J_1(\phi))H_m(J_1(\psi)) + b^2\,\mathcal{E}\,H_m^2(J_1(\psi)).$$

To continue the computations, we need to recall the classical Mehler formula (see, for instance, Bryc 1995, theorem 2.4.1). It says that for a bivariate Gaussian vector (X, Y) with $E(X) = E(Y) = 0$, $E(X^2) = E(Y^2) = 1$ and $E(XY) = \rho$ the joint density f of (X, Y) has the form

$$f(x, y) = \sum_{k=0}^{\infty} \frac{\rho^k}{k!} H_k(x) H_k(y) f_X(x) f_Y(y), \tag{5.1}$$

where $f_X(x) = f_Y(x) = \frac{1}{\sqrt{2\pi}}\exp\left(-\frac{x^2}{2}\right)$ is the marginal density of X and Y. Note that (5.1) yields

$$E(H_m(X)|Y) = \rho^m H_m(Y)$$

for any $m = 1, 2, \ldots$. This is easily visible since for any bounded Borel function g by orthogonality of Hermité polynomials we have

$$E\,E(H_m(X)|Y)g(Y) = E\,H_m(X)g(Y)$$

$$= \int_{\mathbf{R}}\int_{\mathbf{R}} H_m(x)g(y)\left(\sum_{k=0}^{\infty} \frac{\rho^k}{k!} H_k(x) H_k(y) f_X(x) f_Y(y)\right) dx dy$$

$$= \int_{\mathbf{R}}\int_{\mathbf{R}} g(y)\frac{\rho^m}{m!} H_m^2(x) H_m(y) f_X(x) f_Y(y)\, dx dy$$

$$= \int_{\mathbf{R}} \rho^m H_m(y)g(y)f_Y(y)\, dy = \rho^m E\,H_m(Y)g(Y).$$

Thus

$$\mathcal{E}\,J_m^2(a\phi^{\otimes m} + b\psi^{\otimes m}) = a^2\left[E\,\phi^2(X)\right]^m + 2ab\left[E\,\phi(X)\psi(X)\right]^m + b^2\left[E\,\psi^2(X)\right]^m$$

$$= E\left(a\prod_{i=1}^{m}\phi(X_i) + b\prod_{i=1}^{m}\psi^k(X_i)\right)^2 = E\left[\left(a\phi^{\otimes m} + b\psi^{\otimes m}\right)(X_1, \ldots, X_m)\right]^2,$$

where $X_1, \ldots, X_m$ are independent copies of X (with distribution ν).

From the preceding computation it follows that J_m is a linear isometry between the space $\mathcal{T}_s^{(m)}$ and the subspace $J_m\left(\mathcal{T}_s^{(m)}\right)$ of $L^2(\widetilde{\Omega}, \widetilde{\mathcal{F}}, \mathcal{P})$.

In our next step leading to a definition of a multiple Wiener-Itô integral we shall prove that the space $\mathcal{T}_s^{(m)}$ is dense in the symmetric subspace $L_s^2(\mathbf{R}^m, \mathcal{B}_m, \nu_m)$. To this end we need an auxiliary result which will be used in the expansion for symmetric functions given later in Proposition 5.1.1. Recall that S_m denotes the set of all possible permutations of the set $\{1, \ldots, m\}$.

Lemma 5.1.1. *Let* $a_1, \ldots, a_m \in V$, *where* V *is a linear space. Then*

$$2^m \sum_{\sigma \in S_m} a_{\sigma(1)} \otimes \ldots \otimes a_{\sigma(m)} = \sum_{\underline{\varepsilon} \in \{-1,1\}^m} \left(\prod_{i=1}^m \varepsilon_i \right) \left(\sum_{i=1}^m \varepsilon_i a_i \right)^{\otimes m}, \qquad (5.2)$$

Proof. Note that the right hand side of (5.2) can be written as

$$R = \sum_{\underline{\varepsilon} \in \{-1,1\}^m} \left(\prod_{i=1}^m \varepsilon_i \right) \sum_{\underline{i} \in \{1,\ldots,m\}^m} \left(\prod_{r=1}^m \varepsilon_{i_r} \right) (a_{i_1} \otimes \ldots \otimes a_{i_m})$$

$$= \sum_{\underline{i} \in \{1,\ldots,m\}^m} (a_{i_1} \otimes \ldots \otimes a_{i_m}) \sum_{\underline{\varepsilon} \in \{-1,1\}^m} \left(\prod_{i=1}^m \varepsilon_i \right) \left(\prod_{r=1}^m \varepsilon_{i_r} \right)$$

with $\underline{\varepsilon} = (\varepsilon_1, \ldots, \varepsilon_m)$ and $\underline{i} = (i_1, \ldots, i_m)$. But

$$\prod_{r=1}^m \varepsilon_{i_r} = \prod_{i=1}^m \varepsilon_i^{\delta_i},$$

where

$$\delta_i = \#\{j \in \{i_1, \ldots, i_m\} : j = i\}, \qquad i = 1, \ldots, m.$$

Moreover, we have

$$\sum_{\underline{\varepsilon} \in \{-1,1\}^m} \left(\prod_{i=1}^m \varepsilon_i \right) \left(\prod_{i=1}^m \varepsilon_i^{\delta_i} \right) = \sum_{\underline{\varepsilon} \in \{-1,1\}^m} \prod_{i=1}^m \varepsilon_i^{1+\delta_i} = \prod_{i=1}^m \left(\sum_{\varepsilon_i \in \{-1,1\}} \varepsilon_i^{1+\delta_i} \right)$$

$$= \begin{cases} 2^m, & \text{if } \forall i \in \{1, \ldots, m\} \ \delta_i = 1, \\ 0, & \text{otherwise.} \end{cases}$$

Thus

$$R = 2^m \sum_{\underline{i} \in \{1,\ldots,m\}^m} (a_{i_1} \otimes \ldots \otimes a_{i_m}) \, I(\{i_1, \ldots, i_m\} = \{1, \ldots, m\})$$

$$= 2^m \sum_{\sigma \in S_m} a_{\sigma(1)} \otimes \ldots \otimes a_{\sigma(m)}. \qquad \square$$

With the result of the lemma established we may now prove that $\mathcal{T}_s^{(m)}$ is dense in $L_s^2(\mathbf{R}^m, \mathcal{B}_m, \nu_m)$.

Proposition 5.1.1. *There exists a basis* (ψ_r) *in* $L^2(\mathbf{R}, \mathcal{B}, \nu)$ *such that for any* $f \in L_s^2(\mathbf{R}^m, \mathcal{B}_m, \nu_m)$

$$f = \sum_{r=1}^\infty \alpha_r \psi_r^{\otimes m} \qquad (5.3)$$

where (α_r) is a sequence of real numbers such that the double series

$$||f||^2 = \sum_{r,s=1}^{\infty} \alpha_r \alpha_s \rho^m(r,s) < \infty, \tag{5.4}$$

where $\rho(r,s) = E(\psi_r(X)\psi_s(X))$.

Proof. Since f is symmetric we have

$$f(x_1,\ldots,x_m) = \frac{1}{m!} \sum_{\sigma \in \Pi_m} f(x_{\sigma(1)},\ldots,x_{\sigma(m)}).$$

Now, from the theory of Hilbert spaces, it follows that there exists an orthonormal basis (ϕ_r) in $L^2(\mathbf{R},\mathcal{B},\nu)$ such that

$$f(x_1,\ldots,x_m) = \frac{1}{m!} \sum_{\sigma \in S_m} \sum_{\underline{i} \in \mathbf{N}^m} \beta_{\underline{i}} \prod_{l=1}^{m} \phi_{i_l}(x_{\sigma(l)})$$

$$= \frac{1}{m!} \sum_{\underline{i} \in \mathbf{N}^m} \beta_{\underline{i}} \sum_{\sigma \in S_m} \prod_{l=1}^{m} \phi_{i_l}(x_{\sigma(l)})$$

$$= \frac{1}{m!} \left(\sum_{\underline{i} \in \mathbf{N}^m} \beta_{\underline{i}} \sum_{\sigma \in S_m} \phi_{i_{\sigma(1)}} \otimes \ldots \otimes \phi_{i_{\sigma(m)}} \right)(x_1,\ldots,x_m),$$

with $\sum_{\underline{i} \in \mathbf{N}^m} \beta_{\underline{i}}^2 < \infty$. By Lemma 5.1.1 we have

$$f = \frac{1}{2^m m!} \sum_{\underline{i} \in \mathbf{N}^m} \beta_{\underline{i}} \sum_{\underline{\varepsilon} \in \{-1,1\}^m} \left(\prod_{l=1}^{m} \varepsilon_l \right) \left(\sum_{l=1}^{m} \varepsilon_l \phi_{i_l} \right)^{\otimes m}$$

$$= \sum_{\underline{i} \in \mathbf{N}^m, \underline{\varepsilon} \in \{-1,1\}^m} \frac{\beta_{\underline{i}} \prod_{l=1}^{m} \varepsilon_l}{2^m m!} \left(\sum_{l=1}^{m} \varepsilon_l \phi_{i_l} \right)^{\otimes m}.$$

Define

$$c_m(\underline{i},\underline{\varepsilon}) = \left\| \sum_{l=1}^{m} \varepsilon_l \phi_{i_l} \right\|^2 = m + 2 \sum_{1 \leq r < s \leq m} \varepsilon_i \varepsilon_j \delta(i_r = i_s).$$

Then we have

$$f = \sum_{\underline{i} \in \mathbf{N}^m, \underline{\varepsilon} \in \{-1,1\}^m} \alpha_{(\underline{i},\underline{\varepsilon})} \, \psi_{(\underline{i},\underline{\varepsilon})}^{\otimes m},$$

where

$$\alpha_{(\underline{i},\underline{\varepsilon})} = \frac{\sqrt{c_m(\underline{i},\underline{\varepsilon})} \beta_{\underline{i}} \prod_{l=1}^{m} \varepsilon_l}{2^m m!} \quad \text{and} \quad \psi_{(\underline{i},\underline{\varepsilon})} = \frac{1}{\sqrt{c_m(\underline{i},\underline{\varepsilon})}} \sum_{l=1}^{m} \varepsilon_l \phi_{i_l}.$$

Note that $(\psi_{(\underline{i},\underline{\varepsilon})})_{(\underline{i},\underline{\varepsilon})}$ is a basis in $L^2(\mathbf{R},\mathcal{B},\nu)$ and the double series of $(\alpha_{(\underline{i},\underline{\varepsilon})})_{(\underline{i},\underline{\varepsilon})}$ is summable, that is (5.4) holds. $\square$

It is obvious that in the expansion (5.3) the elements of the basis (ψ_k) can be taken as standardized, that is except of $\int_{\mathbf{R}} \psi_k(x)\,\nu(dx) = 0$ also $\int_{\mathbf{R}} \psi_k^2(x)\,\nu(dx) = 1$, $k = 1, 2, \ldots$

In view of the above result the multiple Wiener-Itô integral $J_m(f)$ for any symmetric function from the space $L_s^2(\mathbf{R}^m, \mathcal{B}_m, \nu_m)$ may be now defined through the unique extension of the linear isometry J_m from the space $\mathcal{T}_s^{(m)}$ to the whole symmetric space $L_s^2(\mathbf{R}^m, \mathcal{B}_m, \nu_m)$. Thus, J_m is identified with the unique functional on the Gaussian system $\mathcal{J}$ such that

$$\mathcal{E}(J_m^2(f)) = E(f^2(X_1, \ldots, X_m)) \qquad \text{for any } f \in L_s^2(\mathbf{R}^m, \mathcal{B}_m, \nu_m) \qquad (5.5)$$

Moreover $J_m(f)$ can be expanded as a series of Hermité polynomials.

Proposition 5.1.2. *Let (ϕ_k) be a standardized basis in $L^2(\mathbf{R}, \mathcal{B}(\mathbf{R}), \nu)$ such that for any symmetric function $f \in L^2(\mathbf{R}^m, \mathcal{B}_m, \nu_m)$*

$$f = \sum_{r=1}^{\infty} \alpha_r \phi_r^{\otimes m}.$$

Then

$$J_m(f) = \sum_{r=1}^{\infty} \alpha_r H_m(J_1(\phi_r)), \qquad (5.6)$$

Proof. To prove the result it suffices to show that (5.5) holds for $J_m(f)$ defined by (5.6). But we have

$$m!\,\mathcal{E}(J_m^2(f)) = \sum_{r,s=1}^{\infty} \alpha_r \alpha_s \mathcal{E}[H_m(J_1(\phi_r)) H_m(J_1(\phi_s))] = m! \sum_{r,s=1}^{\infty} \alpha_r \alpha_s \rho^m(r, s),$$

where $\rho(r, s) = E(\phi_r(X)\phi_s(X))$.

On the other hand

$$E(f^2(X_1, \ldots, X_m)) = \sum_{r,s=1}^{\infty} \alpha_r \alpha_s \left(E[\phi_r(X)\phi_s(X)] \right)^m.$$

Now the result follows by the definition of the quantities $\rho(r, s)$. □

5.1.2 Classical Limit Theorems for U-statistics

Let $(X_k)_{k \geq 1}$ be a sequence of independent identically distributed random variables. Let $h \in L_s^{(m)}$ be a symmetric kernel. Consider the corresponding U-statistic

$$U_n^{(m)}(h) = \binom{n}{m}^{-1} [\pi_m^n(h)] (X_1, \ldots, X_n) \qquad (5.7)$$

with its H-decomposition

$$U_n(h) - E\, U_n(h) = \sum_{c=r}^{m} \binom{m}{c}\binom{n}{c}^{-1} [\pi_c^n(g_c)](X_1,\ldots,X_n) \qquad (5.8)$$

where (see (1.11))

$$g_c(x_1,\ldots,x_c) = \sum_{i=1}^{c}(-1)^{c-i} \sum_{1\le j_1<\ldots<j_i\le c} \tilde{h}_i(x_{j_1},\ldots,x_{j_i}). \qquad (5.9)$$

with $\tilde{h}_c = h_c - E(h)$ and

$$h_c(x_1,\ldots,x_c) = E(h(x_1,\ldots,x_c,X_{c+1},\ldots,X_m))$$

for $c = 1,2,\ldots,m$. We may write the complete degeneracy property of g_c (1.12) more concisely as

$$E\, g_c(x_1,\ldots,x_{c-1},X_c) = 0, \qquad c = 1,2,\ldots,m, \qquad (5.10)$$

for any $x_1,\ldots,x_m$.

Recall that the number $r - 1$ in (5.8) is the degeneration (or degeneracy) level of the U-statistic where $r = \min\{c \ge 1 : h_c \not\equiv 0\}$. In what follows we also refer to r as non-degeneracy level. It appears that the non-degeneracy level is essential for the limiting behavior of U-statistics.

For $r = 1$ it was proved by Hoeffding (1948) that

$$\sqrt{n}[U_n(h) - E(U_n(h))] \xrightarrow{d} mN(0, Eg_1^2(Y_1)).$$

The case of $r = 2$ waited for over three decades, until Serfling (1980) showed that

$$n(U_n(h) - E(U_n(h))) \xrightarrow{d} \binom{m}{2}\sum_{k=1}^{\infty}\lambda_k(\mathcal{Z}_k - 1),$$

where $(\mathcal{Z}_k)$ is a sequence of independent chi-square with one degree of freedom random variables, and (λ_k) are defined by the decomposition of g_2:

$$g_2(x_1, x_2) = \sum_{k=1}^{\infty}\lambda_k\phi_k(x_1)\phi_k(x_2),$$

where (ϕ_k) is an orthonormal basis in $L^2(\mathbf{R}, \mathcal{B}, \nu)$, ν being the common distribution of X_i's. Dynkin and Mandelbaum (1983) considered the case of an arbitrary $r \ge 1$. They showed that

$$\binom{m}{r}^{-1} n^{r/2}(U_n(h) - E(U_n(h)) \xrightarrow{d} J_r(g_r), \qquad (5.11)$$

where J_r is the r-th multiple Wiener-Itô integral as defined in in the previous section.

5.1.3 Dynkin-Mandelbaum Theorem

This section is devoted to the detailed proof of the Dynkin and Mandelbaum theorem. The proof we offer is more elementary than the original one.

Theorem 5.1.1. *Let (X_n) be a sequence of iid rv's. Let U_n, $n = m, m+1, \ldots$, be a U-statistic for the sequence (X_n) defined by (5.7) (we assume that m is fixed for all n's). Assume that the non-degeneracy level of the kernel h is r and that $E(g_r^2) < \infty$. Then the convergence (5.11) holds.*

Proof. Note that by the representation (5.8) we have

$$\binom{m}{r}^{-1} n^{r/2} \left(U_n - E(U_n)\right)$$

$$= \binom{m}{r}^{-1} n^{r/2} \sum_{k=r+1}^{m} \binom{m}{k}\binom{n}{k}^{-1} \pi_k^n(g_k) + n^{r/2} \binom{n}{r}^{-1} \pi_r^n(g_r).$$

The first term, we call it R_n, in the above sum converges in probability to zero. This is due to the following computation of its variance based on the orthogonality of g_k's and the property (5.10)

$$Var(R_n) = \binom{m}{r}^{-2} n^r \sum_{k=c+1}^{m} \binom{m}{k}^2 \binom{n}{k}^{-2} Var(\pi_k^n(g_k))$$

$$= \binom{m}{r}^{-2} n^r \sum_{k=r+1}^{m} \binom{n}{k}^{-1} E(g_k^2).$$

Due to the obvious inequality $n(n-1)\ldots(n-k+1) > (n/2)^{r+1}$ valid for $k > r$ and n sufficiently large we have

$$Var(R_n) \leq \frac{A}{n}$$

with

$$A = 2^{r+1} \binom{m}{r}^{-2} \sum_{k=r+1}^{m} \binom{m}{k}^2 k! E(g_k^2).$$

Thus we have to show that

$$n^{r/2} \binom{n}{r}^{-1} \pi_r^n(g_r) \xrightarrow{d} J_r(g_r).$$

Note that we can expand g_r as in Proposition 5.1.1, i.e.

$$g_r(x_1, \ldots, x_r) = \sum_{k=1}^{\infty} \alpha_k \psi_k(x_1) \ldots \psi_k(x_r),$$

where $E(\psi_k(X_i)) = 0$ and $E(\psi_k^2(X_i)) = 1$, $k = 1, 2, \ldots$. Moreover

$$\sum_{j,k=1}^{\infty} \alpha_j \alpha_k \rho^r(j,k) = E(g_r^2) < \infty, \tag{5.12}$$

with $\rho(j,k) = \int \psi_j(x)\psi_k(x)\,dP(x)$ - see Prop. 5.1.1. Now, for any K we have

$$n^r \binom{n}{r}^{-2} E\left(\sum_{k=K+1}^{\infty} \alpha_k \sum_{1 \le j_1 < \ldots < j_r \le n} \psi_k(X_{j_1}) \ldots \psi_k(X_{j_r}) \right)^2$$

$$= n^r \binom{n}{r}^{-1} \sum_{j,k=K+1}^{\infty} \alpha_j \alpha_k \rho^r(j,k).$$

Since the first term is bounded uniformly with respect to n then by (5.12) it follows that the above quantity converges to zero as $K \to \infty$ uniformly in n. Hence

$$n^{r/2} \binom{n}{r}^{-1} \sum_{k=K+1}^{\infty} \alpha_k \sum_{1 \le j_1 < \ldots < j_r \le n} \psi_k(X_{j_1}) \ldots \psi_k(X_{j_r}) \xrightarrow{P} 0$$

as $K \to \infty$ uniformly in n. Thus to study the limiting behaviour of $n^{r/2} \binom{n}{r}^{-1} \pi_r^n(g_r)$ it suffices to consider the finite sum

$$n^{-r/2} r! \sum_{k=1}^{K} \alpha_k \sum_{1 \le j_1 < \ldots < j_r \le n} \psi_k(X_{j_1}) \ldots \psi_k(X_{j_r}).$$

Note that the inner sum is the elementary symmetric polynomial $S_n(r)$ in variables $\psi_k(X_1), \ldots, \psi_k(X_n)$, which will be denoted by $S_n^{(k)}(r)$.

Due to the recursion formula for the elementary symmetric polynomials (see Lemma 3.2.1) which can be rewritten as

$$cS_n^{(k)}(c) = \sum_{d=0}^{c-1}(-1)^d S_n^{(k)}(c - d - 1) \sum_{j=1}^{n} \psi_k^{d+1}(X_j), \qquad k = 1, \ldots, K, \tag{5.13}$$

it follows that for any $c = 1, 2, \ldots$ there exists a function F_c of c variables such that for any n

$$c! n^{-c/2} S_n^{(k)}(c) = F_c\left(n^{-j/2} \sum_{i=1}^{n} \psi_k^j(X_i),\ j = 1, \ldots, c \right), \qquad k = 1, \ldots, K.$$

Moreover, it follows from (5.13) that

$$F_c(x, 1, 0, \ldots, 0) = xF_{c-1}(x, 1, 0, \ldots, 0) - (c-1)F_{r-2}(x, 1, 0, \ldots, 0), \qquad c = 1, 2, \ldots,$$

with $F_{-1}(x, 1, 0, \ldots, 0) = 0$ and $F_0(x, 1, 0, \ldots, 0) = 1$. Consequently, see the recurrence for Hermité polynomials given in (3.7), $F_c(x, 1, 0, \ldots, 0)$ is a monic Hermité polynomial H_c, $c = 0, 1, \ldots$.

Note that, by the classical central limit theorem the weak convergence

$$\left(\frac{1}{\sqrt{n}} \sum_{i=1}^{n} \psi_k(X_i), \ k = 1, \ldots, K \right) \xrightarrow{d} (J_1(\psi_k), \ k = 1, \ldots, K)$$

holds for any K. Additionally, by the weak law of large numbers, it follows that

$$\left(\frac{1}{n} \sum_{i=1}^{n} \psi_k^2(X_i), \ k = 1, \ldots, K \right) \xrightarrow{P} (1, \ldots, 1)$$

and for $j > 2$ (see the weak law of large numbers - Theorem 3.2.4)

$$\left(\frac{1}{n^{j/2}} \sum_{i=1}^{n} \psi_k^j(X_i), \ k = 1, \ldots, K \right) \xrightarrow{P} (0, \ldots, 0).$$

Consequently, since F_r is continuous, it further follows that

$$n^{-r/2} r! \sum_{k=1}^{K} \alpha_k \sum_{1 \leq j_1 < \ldots < j_r \leq n} \psi_k(X_{j_1}) \ldots \psi_k(X_{j_r}) \xrightarrow{d} \sum_{k=1}^{K} \alpha_k H_r(J_1(\psi_k)).$$

Note that, by (5.12) the variance of the tail

$$Var \left(\sum_{k=K+1}^{\infty} \alpha_k H_r(J_1(\psi_k)) \right) = \sum_{j,k=K+1}^{\infty} \alpha_j \alpha_k \rho^r(j, k) \to 0$$

as $K \to \infty$. Thus the sequence $\left(\sum_{k=1}^{K} \alpha_k H_r(J_1(\psi_k)) \right)$ converges in probability to $J_r(g_r)$ as $K \to \infty$. □

5.1.4 Limit Theorem for U-statistics of Increasing Order

If the order m of the kernel $h = h_m$ of the U-statistic $U_n(h) = U_n^{(m)}$ increases with $n \to \infty$ in such a way that $m/\sqrt{n} \to \lambda \geq 0$ then under certain assumptions on the elements of the Hoeffding decomposition $g_{m,c}$ the limiting distribution of $U_n^{(m)}$ as $n \to \infty$ can be represented as the distribution of a multiple Wiener-Itô integral (in the case $\lambda = 0$) or as an infinite sum of multiple Wiener-Itô integrals (in the case $\lambda > 0$). Such results were originally stated by Korolyuk and Borovskikh (1990). Their derivations were based upon reducing the problem to the main theorem of Dynkin and Mandelbaum (1983) paper, which describes the limit of an infinite series of normalized elements of the Hoeffding decomposition. Below we shall prove analogous results in a simpler way, exploring the techniques which were developed in earlier chapters.

We start with the result for $\lambda = 0$. Here the proof borrows a lot from the proof of the Dynkin-Mandelbaum theorem of the previous section.

Theorem 5.1.2. *Let (X_i) be a sequence of iid rv's. Let $(U_n^{(m_n)})$ be a sequence of U-statistics defined for (X_i) such that they have the same non-degeneracy level r. Assume that $m_n^2/n \to \lambda > 0$. Assume that $Eh_{m_n}^2 < \infty$ $n = 1, 2, \ldots$ and that $E(g_{m_n, r} - g_r)^2 \to 0$ as $n \to \infty$ for a function $g_r : \mathbf{R}^r \to \mathbf{R}$ satisfying (5.10). Let*

$$\sum_{k=r+1}^{m_n} \frac{\left(\frac{m_n^2}{n}\right)^{k-r}}{k!} E\, g_{m_n, k}^2 \to 0. \qquad (5.14)$$

Then

$$\binom{m_n}{r}^{-1} n^{r/2} \left[U_n^{(m_n)} - E\left(U_n^{(m_n)} \right) \right] \xrightarrow{d} J_r(g_r).$$

Proof. As in the proof of Theorem 5.1.1 we decompose the normalized U-statistic as

$$\binom{m}{r}^{-1} n^{r/2} \left(U_n^{(k)} - E(U_n^{(k)}) \right) = R_n + n^{r/2} \binom{n}{r}^{-1} \pi_r^n(g_{m_n, r}),$$

where

$$R_n = \binom{m_n}{r}^{-1} n^{r/2} \sum_{k=r+1}^{m_n} \binom{m_n}{k} \binom{n}{k}^{-1} \pi_k^n(g_{m_n, k}).$$

Note that

$$Var(R_n) = \binom{m_n}{r}^{-2} n^r \sum_{k=r+1}^{m_n} \binom{m_n}{k}^2 \binom{n}{k}^{-1} E\, g_{m_n, k}^2$$

as well as, for any $k = r+1, \ldots, m_n$

$$\binom{m_n}{r}^{-2} \binom{m_n}{k}^2 \leq \left(\frac{r!}{k!} \right)^2 m_n^{2(k-c)}$$

and

$$n^r \binom{n}{k}^{-1} \leq \left(\frac{n}{n-r+1} \right)^r \frac{1}{(n-r)\ldots(n-k+1)}$$

$$\leq \left(\frac{n}{n-r+1} \right)^r \left(\frac{n}{n-m_n} \right)^{m_n} \frac{1}{n^{k-r}}.$$

Since the first two terms on the right hand side of the last inequality are bounded (each converges to 1), then it follows that

$$\binom{m_n}{r}^{-2} n^r \binom{m_n}{k}^2 \binom{n}{k}^{-1} \leq C \frac{(m_n^2/n)^{k-r}}{k!},$$

where C is a constant. Consequently, by (5.14) it follows that $Var(R_n) \to 0$.

Since, as in the proof of Theorem 5.1.1 it follows that

$$n^{r/2} \binom{n}{r}^{-1} \pi_r^n(g_r) \xrightarrow{d} J_r(g_r),$$

it suffices to show that

$$n^{r/2} \binom{n}{r}^{-1} (\pi_r^n(g_{m_n,r}) - \pi_r^n(g_r)) \xrightarrow{P} 0.$$

This follows from the fact that the second moment of this difference is

$$n^r \binom{n}{r}^{-1} E \left(g_{m_n,r} - g_r \right)^2 \leq \left(\frac{n}{n-r+1} \right)^r E \left(g_{m_n,r} - g_r \right)^2$$

and thus, the assumption of the theorem implies that it converges to 0.   □

The analogous result for $\lambda > 0$ is somewhat more difficult. In particular, an infinite series of multiple Wiener-Itô integrals appears in the limit.

Theorem 5.1.3. *Let (X_i) be a sequence of iid rv's. Let $(U_n^{(m_n)})$ be a sequence of U-statistics defined for (X_i) such that they have the same non-degeneracy level r. Assume that $m_n^2/n \to \lambda > 0$. Assume that $Eh_{m_n}^2 < \infty \ n = 1, 2, \ldots,$ and that there exists a sequence of functions $\mathbf{g} = (g_k)$ such that g_k is a symmetric function on $\mathbf{R}^k$ satisfying (5.10), $k = r, r+1, \ldots$. Moreover, let*

$$\sum_{k=r}^{\infty} \frac{\lambda^k}{k!} E \, g_k^2 < \infty \qquad and \qquad \sum_{k=r}^{\infty} \frac{\left(\frac{m_n^2}{n} \right)^k}{k!} E \, g_k^2 < \infty \qquad uniformly \ in \ n \tag{5.15}$$

and

$$\sum_{k=r}^{m_n} \frac{\left(\frac{m_n^2}{n} \right)^k}{k!} E \left(g_{m_n,k} - g_k \right)^2 \to 0. \tag{5.16}$$

Then

$$Z_n = U_n^{(m_n)} - E \left(U_n^{(m_n)} \right) \xrightarrow{d} \sum_{k=r}^{\infty} \frac{\lambda^{k/2}}{k!} J_k(g_k).$$

Proof. Consider first the random variable

$$S_{N,n} = \sum_{k=r}^{N} \binom{m}{k} \binom{n}{k}^{-1} \pi_k^{(n)}(g_k) = \sum_{k=r}^{N} \sum_{j=1}^{\infty} \alpha_{j,k} \binom{m}{k} \binom{n}{k}^{-1} \pi_k^{(n)} \psi_j^{\otimes k},$$

where $g_k = \sum_{j=1}^{\infty} \alpha_{k,j} \psi_j^{\otimes k}$, $k = r, \ldots, N$ and $m = m_n$. Note that the sequence (ψ_j) is common for all k's. This is possible by defining it for the largest space $L^2(\mathbf{R}^N, \mathcal{B}(\mathbf{R}^N), P_X^{\otimes N})$. Moreover, $E(\psi_j(X_i)) = 0$ and $E(\psi_j^2(X_i)) = 1$, $j = 1, 2, \ldots$.

Since the above sum with respect to k is finite, similarly as in the proof of the previous result it follows that

$$\sum_{k=r}^{N}\sum_{j=K+1}^{\infty}\alpha_{j,k}\binom{m}{k}\binom{n}{k}^{-1}\pi_k^{(n)}\psi_j^{\otimes k}\xrightarrow{P}0$$

as $K \to \infty$ uniformly in n. Thus using the fact the $m^2/n \to \lambda$ and properties of elementary symmetric polynomials as in the previous proof we get

$$\sum_{k=r}^{N}\sum_{j=1}^{K}\alpha_{j,k}\left(\frac{m}{\sqrt{n}}\right)^k\frac{\binom{m}{k}(\sqrt{n})^k}{m^k\binom{n}{k}}\pi_k^{(n)}\psi_j^{\otimes k}\xrightarrow{d}\sum_{k=r}^{N}\sum_{j=1}^{K}\alpha_{j,k}\lambda^{k/2}H_k(J_1(\psi_j))$$

Note that as $K \to \infty$ the right hand side of the above expression converges in probability (see the previous proof) to $\sum_{k=r}^{N}\lambda^{k/2}J_k(g_{k,})$.

Note that by orthogonality (5.10) we have

$$A_n = E\left(\sum_{k=r}^{m}\frac{\binom{m}{k}}{\binom{n}{k}}\pi_k^{(n)}(g_{m,k}-g_k)\right)^2 = \sum_{k=r}^{m}\frac{\binom{m}{k}^2}{\binom{n}{k}}E(g_{m,k}-g_k)^2.$$

Thus the inequality $m^2/n > (m-1)^2/(n-1)$ and the condition (5.16) imply

$$A_n \le \sum_{k=r}^{m}\frac{\left(\frac{m^2}{n}\right)^k}{k!}E\left(g_{m,k}-g_k\right)^2 \to 0.$$

Moreover, the second part of (5.15) implies that

$$\sum_{k=m+1}^{\infty}\frac{\left(\frac{m^2}{n}\right)^k}{k!}Eg_k^2 \to 0.$$

Consequently,

$$\sum_{k=r}^{m}\frac{\binom{m}{k}}{\binom{n}{k}}\pi_k^{(n)}g_{m,k} \qquad \text{and} \qquad \sum_{k=r}^{\infty}\frac{\binom{m}{k}}{\binom{n}{k}}\pi_k^n(g_k)$$

are asymptotically (as $n \to \infty$) equivalent in distribution.

Note also that $\sum_{k=N}^{\infty}\frac{\lambda^{k/2}}{k!}J_k(g_k)$ converges in probability to 0 as $N \to \infty$: by the orthogonality of g_k's and properties of the multiple Wiener-Itô integral (see, Dynkin and Mandelbaum 1983, formula (2.2)) we get

$$E\left(\sum_{k=N}^{\infty}\frac{\lambda^{k/2}}{k!}J_k(g_k)\right)^2 = \sum_{k=N}^{\infty}\frac{\lambda^k}{k!}E\,g_k^2,$$

which by (5.15) converges to 0 with $N \to \infty$.

Finally, we consider

$$T_{N,n} = \sum_{k=N+1}^{m} \frac{\binom{m}{k}}{\binom{n}{k}} \pi_k^n(g_k).$$

Observe that by the orthogonality of g_k's

$$E(T_{N,n}^2) = \sum_{k=N+1}^{m} \frac{\binom{m}{k}^2}{\binom{n}{k}} E(g_k^2) \le \sum_{k=N+1}^{m} \frac{\left(\frac{m^2}{n}\right)^k}{k!} E(g_k^2)$$

and thus by the second part of (5.15) it follows that $T_{N,n}$ converges in probability (uniformly in n) as $N \to \infty$ to 0.

Since for any $\varepsilon > 0$ and any $x \in \mathbf{R}$ we have

$$P(S_{N,n} \le x-\varepsilon)-P(|T_{N,n}| > \varepsilon) \le P(Z_n \le x) \le P(S_{N,n} \le x+\varepsilon)+P(|T_{N,n}| > \varepsilon),$$

the relations which has already been used in Chapter 3, the result follows because of the asymptotic properties of $S_{N,n}$ and $T_{N,n}$ derived in the course of the proof.

5.2 Asymptotics for P-statistics

Let $M_{m \times n}$ be the space of $m \times n$ matrices with real entries and let $h \in L_s^{(m)}$. Further, let $\mathbb{X}^{(m,n)}$ be a random matrix assuming values in $M_{m \times n}$ with iid entries. Let $E(|h^{(m)}|) < \infty$. Then from Theorem 2.2.1 we obtain for the associated generalized permanent function (under slightly different notation)

$$Per_{h^{(m)}}^{(m,n)} \mathbb{X}^{(m,n)} = E\left(Per_{h^{(m)}}^{(m,n)} \mathbb{X}^{(m,n)}\right) + m!\binom{n}{m} \sum_{k=1}^{m} \frac{(n-k)!}{n!} W_{g_{m,k}}^{(m,n)},$$

where

$$W_{g_{m,k}}^{(m,n)} = \sum_{1 \le i_1 < \ldots < i_k \le m} \sum_{1 \le j_1 < \ldots < j_k \le n} \sum_{\sigma \in \Pi_m} g_{m,k}\left(X_{i_{\sigma(1)},j_1}, \ldots, X_{i_{\sigma(k)},j_k}\right)$$

and

$$g_{m,k}(w_1, \ldots, w_c)$$
$$= \int_{\mathbf{R}^m} h^{(m)}(z_1, \ldots, z_m) \left(\prod_{r=1}^{k}(\delta_{w_r}(dz_r) - P_X(dz_r))\right)\left(\prod_{s=k+1}^{m} P_X(dz_s)\right)$$

for P_X being the distribution of $X_{1,1}$. Recall that, similarly as for U-statistics, we say that r is non-degeneracy level of P-statistics if $g_k^{(m)} \equiv 0$ for $k < r$ and $g_r^{(m)} \not\equiv 0$.

The main object of this section is to prove limit theorems for P-statistics with $m_n/n \to \lambda \geq 0$. Similarly, as for U-statistics, two cases $\lambda = 0$ and $\lambda > 0$, differ and they are treated separately. However in both cases the asymptotic behaviour of the sequence of centered and normalized P-statistics

$$\left(\frac{Per_{h(m_n)}^{(m_n,n)}(\mathbb{X}^{(m_n,n)}) - E\left(Per_{h(m_n)}^{(m_n,n)}(\mathbb{X}^{(m_n,n)}) \right)}{\binom{n}{m_n} m_n!} \right)_n$$

is compared to the asymptotics of respective U-statistics.

Again, as for U-statistics, we start with the case $\lambda = 0$.

Theorem 5.2.1. *Let $(\mathbb{X}^{(m_n,n)})$ be a sequence of matrices in $M_{m_n \times n}$ embedded in an infinite matrix $\mathbb{X}^{(\infty)}$ of iid entries. Assume $m_n/n \to 0$. Consider a sequence of P-statistics $\left(Per_{h(m_n)}^{(m_n,n)}(\mathbb{X}^{(m_n,n)}) \right)$ with a common level of non-degeneracy equal to r. Assume that $E\left[(h^{(m_n)})^2 \right] < \infty$, $n = 1, 2, \ldots$, and that*

$$\sum_{k=r+1}^{m_n} \frac{\left(\frac{m_n}{n}\right)^{k-r}}{k!} E\, g_{m_n,k}^2 \to 0. \tag{5.17}$$

Assume also that $E(g_{m_n,r} - g_r)^2 \to 0$ as $n \to \infty$ for a symmetric function $g_r : \mathbf{R}^r \to \mathbf{R}$ satisfying (5.10).

Then

$$r! \left(\frac{n}{m_n} \right)^{r/2} \frac{Per_{h(m_n)}^{(m_n,n)}(\mathbb{X}^{(m_n,n)}) - E\left(Per_{h(m_n)}^{(m_n,n)}(\mathbb{X}^{(m_n,n)}) \right)}{\binom{n}{m_n} m_n!} \xrightarrow{d} J_r(g_r).$$

Proof. Write $m = m_n$. Consider a U-statistic $U_{mn}^{(m)}$ with the kernel h_m based on an iid sample of $m_n n$ random variables: $X_{1,1}, \ldots, X_{m,n}$. Then $m^2/(mn) = m/n \to \lambda$. Consequently, the assumptions of Theorem 5.1.2 are satisfied (with n changed into mn) and thus

$$\binom{m}{r}^{-1} (mn)^{r/2} \left(U_{mn}^{(m_n)} - E(U_{mn}^{(m)}) \right) \xrightarrow{d} J_r(g_r).$$

To complete the proof we will argue that

$$\frac{r! n^{r/2} \left[Per_{h(m)}^{(m,n)}(\mathbb{X}^{(m,n)}) - E\left(Per_{h(m)}^{(m,n)}(\mathbb{X}^{(m,n)}) \right) \right]}{m^{r/2} \binom{n}{m} m!}$$

$$- \frac{(mn)^{r/2} \left[U_{mn}^{(m)} - E(U_{mn}^{(m)}) \right]}{\binom{m}{r}}$$

$$= \Omega_n(\mathbf{g}_m) \xrightarrow{P} 0.$$

Note that symmetry of the kernels $g_{m,k}$'s entails

$$Cov\left((k!)^{-1}\binom{n}{k}^{-1}W_{g_{m,k}}^{(m,n)}, \binom{m}{k}\binom{mn}{k}^{-1}\pi_{mn}^{k}(g_{m,k})\right)$$

$$= Var\left(\binom{m}{k}\binom{mn}{k}^{-1}\pi_{mn}^{k}(g_{m,k})\right) = \frac{\binom{m}{k}^2}{\binom{mn}{k}}E(g_{m,k}^2) \qquad (5.18)$$

for any $k = r, r+1, \ldots, m$. Note also that by (2.13) it follows that

$$Var\left((k!)^{-1}\binom{n}{k}^{-1}W_{g_{m,k}}^{(m,n)}\right) = \frac{\binom{m}{k}}{\binom{n}{k}}\frac{1}{k!}E(g_{m,k}^2). \qquad (5.19)$$

We will show that Ω_n converges to 0 in L_2. By orthogonality of $(g_{m,k})$ it follows that

$$Var\,\Omega_n(\mathbf{g}_m) = \left(\frac{n}{m}\right)^r \sum_{k=r}^{m} Var\left(r!\frac{W_{g_{m,k}}}{k!\binom{n}{k}} - \frac{m^r\binom{m}{k}\pi_{mn}^{k}(g_{m,k})}{\binom{m}{r}\binom{mn}{k}}\right).$$

Thus (5.18) and (5.19) applied to the first element of the above sum yields

$$\left(\frac{n}{m}\right)^r Var\left(r!\frac{W_{g_{m,r}}}{r!\binom{n}{r}} - \frac{m^r\binom{m}{r}\pi_{mn}^{r}(g_{m,r})}{\binom{m}{r}\binom{mn}{r}}\right)$$

$$= \frac{n^r}{m^r}\left[(r!)^2\frac{\binom{m}{r}}{\binom{n}{r}r!} - 2r!\frac{m^r}{\binom{m}{r}}\frac{\binom{m}{r}}{\binom{mn}{r}} + \left(\frac{m^r}{\binom{m}{r}}\right)^2\frac{\binom{m}{r}^2}{\binom{mn}{r}}\right]$$

$$= \frac{\binom{m}{r}}{m^r}\left[\frac{r!n^r}{\binom{n}{r}} - 2\frac{r!(mn)^r}{\binom{mn}{r}} + \frac{m^r}{\binom{m}{r}}\frac{(mn)^r}{\binom{mn}{r}}\right] \qquad (5.20)$$

Since the term in the square brackets above converges to zero and the factor in front of it converges to $r!$ we conclude that the whole quantity converges to zero.

Note that

$$\left(\frac{n}{m}\right)^r \sum_{k=r+1}^{m} Var\left(r!\frac{W_{g_{m,k}}}{k!\binom{n}{k}} - \frac{m^r\binom{m}{k}\pi_{mn}^{k}(g_{m,k})}{\binom{m}{r}\binom{mn}{k}}\right)$$

$$\le \left(\frac{n}{m}\right)^r \sum_{k=r+1}^{m} Var\left(r!\frac{W_{g_{m,k}}}{k!\binom{n}{k}}\right) + \left(\frac{n}{m}\right)^r \sum_{k=r+1}^{m} Var\left(\frac{m^r\binom{m}{k}\pi_{mn}^{k}(g_{m,k})}{\binom{m}{r}\binom{mn}{k}}\right)$$

$$= I_1(n) + I_2(n).$$

Now treat separately $I_1(n)$ and $I_2(n)$.

For $I_1(n)$ we use the second line of (5.19) to get

$$I_1(n) = \frac{(r!)^2 n^r}{m^r} \sum_{k=r+1}^{m} \frac{\binom{m}{k}}{\binom{n}{k}} \frac{1}{k!} E(g_{m,k}^2)$$

$$\leq (r!)^2 \sum_{k=r+1}^{m} \frac{m^{k-r}}{n^{k-r}} \frac{1}{\left(1 - \frac{m}{n}\right)^k} \frac{E(g_{m,k}^2)}{k!}$$

Since for n large enough $\frac{m}{n} < \frac{1}{2}$ then for such n's

$$I_1(n) \leq 2^r (r!)^2 \sum_{k=r+1}^{m} \left(\frac{m}{n}\right)^{k-r} \frac{E(g_{m,k}^2)}{k!}$$

and thus by (5.17) it follows that $I_1(n)$ converges to zero.
For $I_2(n)$ we have

$$I_2(n) = \left(\frac{n}{m}\right)^r \left(\frac{m^r}{\binom{m}{r}}\right)^2 \sum_{k=r+1}^{m} \frac{\binom{m}{k}^2}{\binom{mn}{k}} E\, g_{m,k}^2$$

$$= \left(\frac{m^r}{\binom{m}{r}}\right)^2 \sum_{k=r+1}^{m} \frac{\binom{m}{k}}{m^r} \frac{n^k \binom{m}{k}}{n^{k-r} \binom{mn}{k}} E\, g_{m,k}^2$$

$$\leq \left(\frac{m^r}{\binom{m}{r}}\right)^2 \frac{1}{\left(1 - \frac{1}{n}\right)^m} \sum_{k=r+1}^{m} \frac{\left(\frac{m}{n}\right)^{k-c}}{k!} E\, g_{m,k}^2.$$

Since the expression standing just in front of the sum above is bounded then
(5.17) implies that $I_2(n) \to 0$.

Thus $Var\, \Omega_n(\mathbf{g}_m) \to 0$ implying $\Omega_n(\mathbf{g}_m) \xrightarrow{P} 0$.   □

In our next result we consider the case $\lambda > 0$.

Theorem 5.2.2. *Let* $(\mathbb{X}^{(m_n, n)})$ *be a sequence of matrices in* $M_{m_n \times n}$ *embedded in an infinite matrix* $\mathbb{X}^{(\infty)}$ *of iid entries. Assume* $m_n/n \to \lambda > 0$. *Consider a sequence of P-statistics* $\left(Per_{h^{(m_n)}}^{(m_n, n)}(\mathbb{X}^{(m_n, n)})\right)$ *with a common level of non-degeneracy equal to* r. *Assume that* $E\left[(h^{(m_n)})^2\right] < \infty$, $n = 1, 2, \ldots$, *and that there exists a sequence of functions* $\mathbf{g} = (g_k)$, *such that* g_k *is a symmetric function on* $\mathbf{R}^k$ *satisfying (5.10),* $k = c, c+1, \ldots$ *as well as*

$$\sum_{k=r}^{\infty} \frac{\lambda^k}{k!} E\, g_k^2 < \infty \qquad and \qquad \sum_{k=r}^{\infty} \frac{\left(\frac{m_n}{n}\right)^k}{k!} E\, g_k^2 < \infty \qquad uniformly\ in\ n$$

$$(5.21)$$

and

$$\sum_{k=r}^{m_n} \frac{\left(\frac{m_n}{n}\right)^k}{k!} E\left(g_{m_n,k} - g_k\right)^2 \to 0. \tag{5.22}$$

Then

$$\frac{Per_{h(m_n)}^{(m_n,n)}\left(\mathbb{X}^{(m_n,n)}\right) - E\, Per_{h(m_n)}^{(m_n,n)}\left(\mathbb{X}^{(m_n,n)}\right)}{\binom{n}{m_n}m_n!} \xrightarrow{d} \sum_{k=r}^{\infty} \frac{\lambda^{k/2}}{k!} J_k(g_k).$$

Proof. Consider a U-statistic $U_{m_n n}^{(m_n)}$ with the kernel h_{m_n} based on an iid sample $X_{1,1}, \ldots, X_{m_n,n}$. Then $m_n^2/(m_n n) = m_n/n \to \lambda$. Consequently, the assumptions of Theorem 5.1.3 are satisfied (with n changed into $m_n n$) and thus (writing $m = m_n$)

$$U_{mn}^{(m)} - E\, U_{mn}^{(m)} \xrightarrow{d} \sum_{k=c}^{\infty} \frac{\lambda^{k/2}}{k!} J_k(g_k).$$

To complete the proof we will argue that

$$\frac{1}{\binom{n}{m}m!}\left[Per_{h(m)}^{(m,n)}\left(\mathbb{X}^{(m,n)}\right) - E\, Per_{h(m)}^{(m,n)}\left(\mathbb{X}^{(m,n)}\right)\right] - \left[U_{mn}^{(m)} - E\, U_{mn}^{(m)}\right]$$

$$= \sum_{k=r}^{m}(k!)^{-1}\binom{n}{k}^{-1} W_{g_{m,k}}^{(m,n)}\left(\mathbb{X}^{(m,n)}\right) - \sum_{k=r}^{m}\binom{m}{k}\binom{mn}{k}^{-1}\pi_{mn}^k(g_{m,k})\left(\mathbb{X}^{(m,n)}\right)$$

$$= \Omega_n(\mathbf{g}_m) \xrightarrow{P} 0. \tag{5.23}$$

Firstly, note that by orthogonality of $(g_{m_n,k})_k$ and $(g_k)_k$ and due to (5.22)

$$E\left(\sum_{k=r}^{m}(k!)^{-1}\binom{n}{k}^{-1}\left[W_{g_{m,k}}^{(m,n)} - W_{g_k}^{(m,n)}\right]\right)^2 = \sum_{k=r}^{m}\frac{\binom{m}{k}}{\binom{n}{k}k!}E(g_{m,k} - g_k)^2 \to 0.$$

Similarly, by the inequality

$$\binom{mn}{k} > \binom{m}{k}\binom{n}{k}k! \tag{5.24}$$

it follows that

$$E\left(\sum_{k=r}^{m}\binom{m}{k}\binom{mn}{k}^{-1}\left[\pi_{mn}^k(g_{m,k}) - \pi_{mn}^k(g_k)\right]\right)$$

$$= \sum_{k=r}^{m}\binom{m}{k}^2\binom{mn}{k}^{-1}E(g_{m,k} - g_k)^2$$

$$\leq \sum_{k=r}^{m}\frac{\binom{m}{k}}{\binom{n}{k}k!}E(g_{m,k} - g_k)^2 \to 0.$$

Thus to show the relation (5.23) it suffices to prove that

$$\Omega_n(\mathbf{g}) \xrightarrow{P} 0. \tag{5.25}$$

This will be accomplished by showing that the variance of the respective difference tends to zero. Note that symmetry of the kernel g_k's entails

$$Cov\left(\sum_{k=r}^{m}(k!)^{-1}\binom{n}{k}^{-1}W_{g_k}^{(m,n)}, \sum_{k=r}^{m}\binom{m}{k}\binom{mn}{k}^{-1}\pi_{mn}^{k}(g_k)\right)$$

$$= Var\left(\sum_{k=c}^{m}\binom{m}{k}\binom{mn}{k}^{-1}\pi_{mn}^{k}(g_k)\right) = \sum_{k=r}^{m}\frac{\binom{m}{k}^{2}}{\binom{mn}{k}}E(g_k^{2}).$$

Note also that by (2.13) it follows that

$$Var\left(\sum_{k=r}^{m}(k!)^{-1}\binom{n}{k}^{-1}W_{g_k}^{(m,n)}\right) = \sum_{k=r}^{m}\frac{\binom{m}{k}}{\binom{n}{k}}\frac{1}{k!}E(g_k^{2}).$$

Thus

$$Var\left(\Omega_n(\mathbf{g})\right) = \sum_{k=r}^{m}\left[\frac{\binom{m}{k}}{\binom{n}{k}}\frac{1}{k!} - \frac{\binom{m}{k}^{2}}{\binom{mn}{k}}\right]E(g_k^{2}).$$

Hence, by the second part of (5.21), we have

$$\sum_{k=N}^{m}\frac{\binom{m}{k}}{\binom{n}{k}}\frac{1}{k!}E(g_k^{2}) \to 0$$

uniformly in n as $N \to \infty$. Similarly, using once again the inequality (5.24), we get

$$\sum_{k=N}^{m}\frac{\binom{m}{k}^{2}}{\binom{mn}{k}}E(g_k^{2}) \to 0$$

uniformly in n as $N \to \infty$. Now, fixing N sufficiently large we see that

$$\sum_{k=r}^{N}\left[\frac{\binom{m}{k}}{\binom{n}{k}}\frac{1}{k!} - \frac{\binom{m}{k}^{2}}{\binom{mn}{k}}\right]E(g_k^{2}) \to 0$$

as $n \to \infty$ since both

$$\frac{\binom{m}{k}}{\binom{n}{k}}\frac{1}{k!} \quad \text{and} \quad \frac{\binom{m}{k}^{2}}{\binom{mn}{k}}$$

converge, as $n \to \infty$ to the same limit $\frac{\lambda^k}{k!}$ and k assumes only finite number of values $1, \ldots, N$. Hence (5.25) follows and the proof of the result is complete.

$\square$

Let us note that in practice the assumptions 5.21 and 5.22 of the above theorem may be often replaced by a set of alternative conditions being easier to verify. We outline them in the remarks below.

Remark 5.2.1. Suppose that exist positive constants A, B, C and D such that

$$|g_k| \leq AB^k \qquad \text{for any } k = r, r+1, \ldots \qquad (5.26)$$

and

$$|g_{m_n,k}| \leq CD^k \qquad \text{for any } k = r, r+1, \ldots, m_n \ \ n = 1, 2, \ldots, \qquad (5.27)$$

as well as for any $k = r, r+1, \ldots$, a point-wise convergence

$$g_{m_n,k} \to g_k \qquad \text{as } n \to \infty \qquad (5.28)$$

holds, then the assumptions (5.21) and (5.22) of Theorem 5.2.2 are satisfied.

Remark 5.2.2. Since by (5.9)

$$g_{m_n,k} = \sum_{i=1}^{k} (-1)^{k-i} \pi_i^k (\tilde{h}_{m_n,i}), \qquad (5.29)$$

where

$$\tilde{h}_{m_n,i}(x_1, \ldots, x_i) = E(h_{m_n}(x_1, \ldots, x_i, X_{i+1}, \ldots, X_{m_n})) - E(h_{m_n}),$$

then it follows that if there exist positive constants a and b such that

$$|\tilde{h}_{m_n,i}| < ab^i$$

for any $i = r, r+1, \ldots$ and any $n = 1, 2, \ldots$, then

$$|g_{m_n,k}| \leq \sum_{i=1}^{k} \binom{k}{i} ab^i \leq a(1+b)^k$$

for any $k = r, r+1, \ldots, m_n$ and any $n = 1, 2, \ldots$ and therefore (5.27) is satisfied.

Remark 5.2.3. Assume that there exist positive constants A, B, C and D such that

$$E(g_k^2) \leq AB^k \qquad \text{for any } k = r, r+1, \ldots \qquad (5.30)$$

and

$$E(g_{m_n,k}^2) \leq CD^k \quad \text{for any } k = r, r+1, \ldots, m_n \text{ and any } n = 1, 2, \ldots \quad (5.31)$$

It is worthy to note that due to the formula (1.19) written is our current notation as

$$E(g^2_{m_n,k}) = \sum_{i=1}^{k} (-1)^{k-i} \binom{k}{i} E(\tilde{h}^2_{m_n,i}),$$

it follows that (5.31) is implied by the following condition: there exist positive constants c and d such that

$$E(\tilde{h}^2_{m_n,k}) \leq cd^k \qquad \text{for any } k = r, r+1, \ldots, m_n \text{ and for any } n = 1, 2, \ldots.$$
$$(5.32)$$

It is quite easy to see that (5.30) and (5.31) (or (5.26) and (5.27)) and the point-wise convergence (5.28) imply that the assumptions of Theorem 5.2.2 hold true: The first part of (5.21) is obvious. The second part of (5.21) follows immediately from the fact that the sequence (m_n/n) is bounded. To see that (5.22) is also satisfied, let us for any $\varepsilon > 0$ choose M large enough to have

$$\sum_{k=M+1}^{\infty} \frac{\left(\frac{m_n}{n}\right)^k}{k!} \left(AB^k + CD^k\right) < \varepsilon/2.$$

Note that M does not depend on n. Then, by (5.28), we can take N large enough to have for $n \geq N$

$$\sum_{k=r}^{M} \frac{\left(\frac{m_n}{n}\right)^k}{k!} E(g_{m_n,k} - g_k)^2) < \varepsilon/2.$$

Remark 5.2.4. Note that (5.31) implies

$$\sum_{k=r+1}^{m_n} \frac{\left(\frac{m_n}{n}\right)^{k-r}}{k!} E\, g^2_{m_n,k} \leq \frac{m_n}{n} CD^{r+1} \sum_{l=0}^{\infty} \frac{\left(\frac{Dm_n}{n}\right)^l}{l!} = CD^{r+1} \frac{m_n}{n} e^{\frac{Dm_n}{n}} \to 0.$$

Thus (5.17) is satisfied and Theorem 5.2.1 holds. Similarly from Remark 5.2.3 it follows that (5.32) implies (5.17).

5.3 Examples

In this section we shall illustrate the applicability of the results from the previous section by revisiting some of the examples introduced in Chapter 1. We shall simplify slightly the notation by writing $Per\,(h_m)$ for $Per^{(m_n,n)}_{h(m_n)}\mathbb{X}^{(m,n)}$ also m for m_n while continuing to assume that, as in Theorem 5.2.2, $m/n \to \lambda > 0$.

Example 5.3.1 (Number of monochromatic matchings in multicolored graph).
In the setting of Example 1.7.2, let as before $P(X = k) = p_k$, $k = 0, 1, \ldots, N$,
where $N \leq \infty$ and $X_{i,j} \overset{d}{=} X$. If

$$h_m(x_1, \ldots, x_m) = \sum_{k=1}^{N} \prod_{i=1}^{m} I(x_i = k).$$

Then $Per_{h_{m,n}} = \mathcal{M}(m, n)$ counts monochromatic perfect matchings in a corresponding random bipartite graph $(G, [y_{i,j}])$.

Note that $E(h_m) = \sum_{k=1}^{N} p_k^m$ and, moreover,

$$g_{m,k}(x_1, \ldots, x_k) = \sum_{i=1}^{N} p_i^m \prod_{l=1}^{k} \left(\frac{1}{p_i} I(x_l = i) - 1 \right).$$

Consider first the case of $N < \infty$ and $p_i = p$, $i = 1, \ldots, N$. Then $E(h_m) = Np^m$ and changing h_m into $\tilde{h}_m = h_m/(Np^m)$ we get

$$g_{m,k}(x_1, \ldots, x_k)$$
$$= g_k(x_1, \ldots, x_k) = \frac{1}{N} \sum_{i=1}^{N} \prod_{l=1}^{k} \left(\frac{1}{p} I(x_l = i) - 1 \right)$$
$$= \frac{1}{N} \left(\frac{1-p}{p} \right)^{k/2} \sum_{i=1}^{N} \phi_i^{\otimes k}(x_1, \ldots, x_k) ,$$

where

$$\phi_i(x) = \frac{I(x = i) - p}{\sqrt{p(1 - p)}} , \qquad i = 1, \ldots, N,$$

are standardized. Thus if $m/n \to \lambda > 0$ then from Theorem 5.2.2 we conclude that

$$\frac{\mathcal{M}(m, n)}{Np^m \binom{n}{m} m!} \overset{d}{\to} 1 + \frac{1}{N} \sum_{k=1}^{\infty} \frac{\left(\lambda \frac{1-p}{p} \right)^{k/2}}{k!} \sum_{i=1}^{N} H_k(J_1(\phi_i))$$
$$= \frac{1}{N} \exp\left(-\frac{\lambda(1 - p)}{2p} \right) \sum_{i=1}^{N} \exp\left(\sqrt{\lambda} Z_i \right),$$

where $(Z_1, \ldots, Z_N)$ is an N-variate centered normal random vector with covariances: $E(Z_i Z_j) = -1$ for $i \neq j$ and $EZ_i^2 = (1 - p)p^{-1}$, $i, j \in \{1, \ldots, N\}$.

If $m/n \to 0$ then Remark 5.2.4 and Theorem 5.2.1 imply

$$\sqrt{\frac{n}{m}} \left(\frac{M(m, n)}{p^m \binom{n}{m} m!} - N \right) \overset{d}{\to} \mathcal{Z},$$

where $\mathcal{Z}$ is a zero mean Gaussian random variable with variance $Var(\mathcal{Z}) = N^2 \frac{1-p}{p}$.

In the general case, i.e. $N \leq \infty$, let $p = \max_{1 \leq i \leq N} p_i$ and let $j_1, \ldots, j_K$ be such that $p_{j_s} = p$, $s = 1, \ldots, K$. Then we define $\tilde{h}_m = h_m/(Kp^m)$ and note that $E\tilde{h}_m \to 1$ as $n \to \infty$ since

$$\left| E\tilde{h}_m - 1 \right| = \left| \frac{\sum_{i \geq 1} p_i^m}{Kp^m} - 1 \right| = \sum_{i \notin \{j_1, \ldots, j_K\}} \left(\frac{p_i}{p} \right)^m$$

$$\leq \left(\frac{\sup_{i \notin \{j_1, \ldots, j_K\}} p_i}{p} \right)^{m-1} \sum_{i \notin \{j_1, \ldots, j_K\}} \frac{p_i}{p} \to 0$$

on noting that $\sup_{i \notin \{j_1, \ldots, j_K\}} p_i < p$. Consequently,

$$g_{m,k}(x_1, \ldots, x_k) = \frac{1}{K} \sum_{i=1}^{N} \left(\frac{p_i}{p} \right)^m \prod_{l=1}^{k} \left(\frac{1}{p_i} I(x_l = i) - 1 \right)$$

Define

$$g_k(x_1, \ldots, x_k) = \frac{1}{K} \sum_{s=1}^{K} \prod_{l=1}^{k} \left(\frac{1}{p} I(x_l = j_s) - 1 \right)$$

$$= \frac{1}{K} \left(\frac{1-p}{p} \right)^{k/2} \sum_{s=1}^{K} \psi_s^{\otimes k}(x_1, \ldots, x_k),$$

where $\psi_s = \phi_{j_s}$, $s = 1, \ldots, K$, are standardized and $k = 1, 2, \ldots$. Then $|g_k| \leq C^k$ and (5.21) is satisfied. Moreover,

$$|g_{m,k}(x_1, \ldots, x_k) - g_k(x_1, \ldots, x_k)|$$

$$= \frac{1}{K} \sum_{i \notin \{0, j_1, \ldots, j_K\}} \left(\frac{p_i}{p} \right)^m \prod_{l=1}^{k} \left(\frac{1}{p_i} I(x_l = i) - 1 \right)$$

$$\leq C^k \sum_{i \notin \{j_1, \ldots, j_K\}} \left(\frac{p_i}{p} \right)^m \leq C^k \delta^{m-1} \sum_{i \notin \{j_1, \ldots, j_K\}} \frac{p_i}{p} \leq \delta^{m-1} C^k/p \to 0,$$

since $\delta = \sup_{i \notin \{j_1, \ldots, j_K\}} p_i < p$ and thus (5.22) is also satisfied. Consequently, in the case $m/n \to \lambda > 0$ Theorem 5.2.2 implies

$$\frac{\mathcal{M}(m,n)}{Kp^m \binom{n}{m} m!} \xrightarrow{d} 1 + \frac{1}{K} \sum_{k=1}^{\infty} \frac{\left(\lambda \frac{1-p}{p} \right)^{k/2}}{k!} \sum_{i=1}^{K} H_k(J_1(\phi_{j_i}))$$

$$= \frac{1}{K} \exp \left(-\frac{\lambda(1-p)}{2p} \right) \sum_{i=1}^{K} \exp \left(\sqrt{\lambda} Z_i \right),$$

where $(Z_1, \ldots, Z_K)$ is a K-dimensional centered normal random vector with covariances $E(Z_i Z_j) = -1$ for $i \neq j$ and $E(Z_i^2) = (1-p)p^{-1}$, $i, j \in \{1, \ldots, K\}$.

Alternatively, if $m/n \to 0$ then it follows from Theorem 5.2.1 (via Remark 5.2.4) that

$$\sqrt{\frac{n}{m}} \left(\frac{M(m,n)}{p^m \binom{n}{m} m!} - K \right) \xrightarrow{d} \mathcal{Z},$$

where $\mathcal{Z}$ is a zero mean Gaussian random variable with variance $Var(\mathcal{Z}) = K^2 \frac{1-p}{p}$.

In our remaining examples we consider always a set of weights in a complete bipartite graph $K(m,n)$.

Example 5.3.2 (Number of matchings with L red edges in bicolored graph). Consider a random bicolored complete bipartite graph $K(m,n)$ from Example 1.7.3 with edges which are either red ($X_{i,j} = 1$) with probability p or black ($X_{i,j} = 0$) with probability $1 - p$. We are interested in the number $\mathcal{K}(n, \alpha)$ of perfect matchings with a given number L_n of red edges as $n \to \infty$, assuming that $L_n/m \to \alpha \in [0,1]$. Thus $\mathcal{K}(n, \alpha) = Per(h_m)$ for

$$h_m(x_1, \ldots, x_m) = I(x_1 + \ldots + x_m = L_n), \qquad x_j \in \{0, 1\}.$$

Note that

$$E(h_m) = \binom{m}{L_n} p^{L_n} (1 - p)^{m - L_n}.$$

Moreover,

$$h_{m,i}(x_1, \ldots, x_i)$$
$$= \binom{m-i}{L_n - x_1 - \ldots - x_i} p^{L_n - x_1 - \ldots - x_i} (1-p)^{m-i-L_n+x_1+\ldots+x_i}, \qquad x_j \in \{0, 1\}.$$

Consequently,

$$\tilde{h}_{m,i}(x_1, \ldots, x_i)$$
$$= \frac{h_{m,i}(x_1, \ldots, x_i) - E(h_m)}{E(h_m)} = \frac{\binom{m-i}{L_n - x_1 - \ldots - x_i}}{\binom{m}{L_n}} \left(\frac{1-p}{p} \right)^{x_1 + \ldots + x_i} (1-p)^{-i} - 1.$$

and thus by (5.9) we have

$$g_{m,k}(x_1, \ldots, x_k) = \sum_{i=1}^{k} (-1)^{k-i} \sum_{1 \le j_1 < \ldots < j_i \le k} \tilde{h}_{m,i}(x_{j_1}, \ldots, x_{j_i})$$
$$= \sum_{i=1}^{k} (-1)^{k-i} \left(\sum_{1 \le j_1 < \ldots < j_i \le k} \frac{\binom{m-i}{L_n - x_{j_1} - \ldots - x_{j_i}}}{\binom{m}{L_n}} \left(\frac{1-p}{p} \right)^{x_{j_1} + \ldots + x_{j_i}} (1-p)^{-i} - 1 \right).$$

Since

$$\lim_{n \to \infty} \frac{\binom{m-i}{L_n - j}}{\binom{m}{L_n}} = \alpha^j (1 - \alpha)^{i-j}$$

then

$$g_{m,k}(x_1, \ldots, x_k)$$

$$\to \sum_{i=1}^{k} (-1)^{k-i} \left(\sum_{1 \le j_1 < \ldots < j_i \le k} \left(\frac{\alpha}{p} \right)^{x_{j_1} + \ldots + x_{j_i}} \left(\frac{1 - \alpha}{1 - p} \right)^{i - x_{j_1} - \ldots - x_{j_i}} - 1 \right)$$

$$= \prod_{i=1}^{k} \left[\left(\frac{\alpha}{p} \right)^{x_i} \left(\frac{1 - \alpha}{1 - p} \right)^{1 - x_i} - 1 \right]$$

$$= \left(\frac{|\alpha - p|}{\sqrt{p(1 - p)}} \right)^k \prod_{i=1}^{k} \phi(x_i) = g_k(x_1, \ldots, x_k),$$

with a standardized

$$\phi(x) = \frac{\sqrt{p(1 - p)}}{|\alpha - p|} \left[\left(\frac{\alpha}{p} \right)^x \left(\frac{1 - \alpha}{1 - p} \right)^{1 - x} - 1 \right], \quad x \in \{0, 1\}.$$

Note that for any $k \ge c$

$$|g_k| \le \left(\max \left\{ \left| \frac{\alpha}{p} - 1 \right|, \left| \frac{1 - \alpha}{1 - p} - 1 \right| \right\} \right)^k.$$

Also, denoting $s = x_1 + \ldots + x_i$, we have

$$|\tilde{h}_{m,i}| \le \left(\frac{L_n}{m} \right)^s \left(1 - \frac{L_n}{m_n} \right)^{i-s} \frac{1}{p^s} \frac{1}{(1 - p)^{i-s}} + 1 \le 2 \left(\frac{1}{\min\{p, 1 - p\}} \right)^i.$$

Thus by (5.26), (5.27) and (5.28) we conclude that the assumptions of Theorems 5.2.2 and 5.2.1 are satisfied.

Then if $m/n \to \lambda > 0$

$$\frac{\mathcal{K}(n, \alpha)}{\binom{n}{m_n} m_n! \binom{m_n}{L_n} p^{L_n} (1 - p)^{m - L_n}} \xrightarrow{d} 1 + \sum_{k=1}^{\infty} \frac{\left(\sqrt{\lambda} \frac{|\alpha - p|}{\sqrt{p(1-p)}} \right)^k}{k!} H_k(J_1(\phi))$$

$$= \exp \left(\sqrt{\lambda} \mathcal{Z} - \frac{\lambda(\alpha - p)^2}{2p(1 - p)} \right),$$

where $\mathcal{Z}$ is a Gaussian random variable with zero mean and variance $Var(Z) = \frac{(\alpha-p)^2}{2p(1-p)}$.

If $m/n \to 0$ then

$$\sqrt{\frac{n}{m}} \left(\frac{\mathcal{K}(n, \alpha)}{\binom{n}{m_n} m_n! \binom{m_n}{L_n} p^{L_n} (1 - p)^{m - L_n}} - 1 \right) \xrightarrow{d} \mathcal{Z}.$$

Example 5.3.3 (Number of heavy normal matchings). Consider a complete graph $K(m,n)$ with random weights $\mathbb{X}^{(m,n)} = [X_{i,j}]$ which are iid $\mathcal{N}(0,1)$ random variables. We are interested in asymptotics of the number $\mathcal{W}(n,\alpha)$ of perfect matchings for which the average total weight exceeds given level $\alpha > 0$. More precisely, let

$$h_m(x_1,\ldots,x_m) = I(x_1 + \ldots + x_m > \alpha m).$$

Then

$$W_n(\alpha) = Per(h_m).$$

Note that

$$E(h_m) = P(X_1 + \ldots + X_m \geq \alpha m),$$

where (X_i) are iid $\mathcal{N}(0,1)$ random variables. Thus

$$E(h_m) = 1 - \Phi(\alpha\sqrt{m}),$$

where Φ denotes the distribution function of the standard normal distribution. Similarly

$$h_{m,i}(x_1,\ldots,x_i) = P(X_{i+1} + \ldots + X_m \geq \alpha m - s) = 1 - \Phi\left(\frac{\alpha m - s}{\sqrt{m - i}}\right),$$

where $s = x_1 + \ldots + x_i$. Thus

$$\tilde{h}_{m,i} = \frac{1 - \Phi\left(\frac{\alpha m - s}{\sqrt{m-i}}\right)}{1 - \Phi(\alpha\sqrt{m})} - 1, \qquad i = 1,\ldots,m-1. \tag{5.33}$$

Now we use the classical double inequality for the tail of the distribution function of the standard normal distribution (see for instance Feller 1968, chapter 7).

$$\frac{1}{\sqrt{2\pi}} e^{-\frac{x^2}{2}} \left(\frac{1}{x} - \frac{1}{x^3}\right) < 1 - \Phi(x) < \frac{1}{\sqrt{2\pi}} e^{-\frac{x^2}{2}} \frac{1}{x} \tag{5.34}$$

for any $x > 0$.

Since for large n the arguments of Φ in (5.33) are positive we can use (5.34) respectively to the numerator and denominator in (5.33) to get the double inequality

$$\exp\left[-\frac{1}{2}\left(\frac{(\alpha m - s)^2}{m - i} - \alpha^2 m\right)\right] \frac{\alpha\sqrt{m(m-i)}\,[(\alpha m - s)^2 - m + i]}{(\alpha m - s)^3}$$

$$< \tilde{h}_{m,i} + 1 < \exp\left[-\frac{1}{2}\left(\frac{(\alpha m - s)^2}{m - i} - \alpha^2 m\right)\right] \frac{\alpha^3 m^{3/2}\sqrt{m - i}}{(\alpha m - s)(\alpha^2 m - 1)}. \tag{5.35}$$

Passing to the limit as $n \to \infty$ in (5.35) we get

$$\tilde{h}_{m,i} \to e^{\alpha s - \frac{i\alpha^2}{2}} - 1.$$

Consequently, by (5.9)

$$g_{m,k}(x_1, \ldots, x_k)$$

$$= \sum_{i=1}^{k} (-1)^{k-i} \sum_{1 \le j_1 < \ldots < j_i \le k} \tilde{h}_{m,i}(x_{j_1}, \ldots, x_{j_i})$$

$$\to \sum_{i=1}^{k} (-1)^{k-i} \sum_{1 \le j_1 < \ldots < j_i \le k} \left(e^{\alpha(x_{j_1} + \ldots + x_{j_i}) - \frac{i\alpha^2}{2}} - 1 \right)$$

$$= \prod_{i=1}^{k} \left(e^{\alpha x_i - \frac{\alpha^2}{2}} - 1 \right) = \left(e^{\alpha^2} - 1 \right)^{k/2} \prod_{i=1}^{k} \phi^{\otimes k}(x_1, \ldots, x_k) = g_k(x_1, \ldots, x_k),$$

where

$$\phi(x) = \frac{e^{\alpha x - \frac{\alpha^2}{2}} - 1}{\sqrt{e^{\alpha^2} - 1}}$$

is standardized. Note that

$$E(g_k^2) = \left[E \left(e^{\alpha \mathcal{N} - \frac{\alpha^2}{2}} - 1 \right)^2 \right]^k.$$

and thus (5.21) is satisfied.

Since $\tilde{h}_{m,i}$ is not bounded we will use Remark 5.2.3. We need to show (5.32).
Note that

$$E(\tilde{h}_{m,i}^2) = \frac{E\left[\Phi(\alpha\sqrt{m}) - \Phi\left(\frac{\alpha m - S_i}{\sqrt{m-i}} \right) \right]^2}{(1 - \Phi(\alpha\sqrt{m}))^2},$$

where $S_i \sim \mathcal{N}(0, i)$, $i = 1, \ldots, m-1$. The numerator is bounded as follows

$$E\left[\Phi(\alpha\sqrt{m}) - \Phi\left(\frac{\alpha m - S_i}{\sqrt{m-i}} \right) \right]^2$$

$$\le \frac{1}{2\pi} E \left(e^{- \min\left\{ \alpha^2 m, \frac{(\alpha m - S_i)^2}{m-i} \right\}} \left(\alpha\sqrt{m} - \frac{\alpha m - S_i}{\sqrt{m-i}} \right)^2 \right)$$

$$\le \frac{1}{2\pi} \left[e^{-\alpha^2 m} E \left(\alpha\sqrt{m} + \frac{S_i - \alpha m}{\sqrt{m-i}} \right)^2 \right.$$

$$\left. + E \left(e^{- \frac{(S_i - \alpha m)^2}{m-i}} \left(\alpha\sqrt{m} + \frac{S_i - \alpha m}{\sqrt{m-i}} \right)^2 \right) \right].$$

Since $Z = \frac{S_i - \alpha m}{\sqrt{m-i}} \sim \mathcal{N}(-\frac{\alpha m}{\sqrt{m-i}}, \frac{i}{m-i}) = \mathcal{N}(\nu, \sigma^2)$ then

$$E(Z + \alpha\sqrt{m})^2 = \sigma^2 + \nu^2 + 2\alpha\sqrt{m}\nu + \alpha^2 m = \frac{i}{m-i}\left(1 + \frac{\alpha^2 m}{\sqrt{m} + \sqrt{m-i}}\right)$$

$$\leq (m-1)(1 + \alpha^2\sqrt{m}) \leq (m-1)(1 + 2\alpha^2 m)e^{\alpha^2 i}$$

where the particular form of the last inequality is convenient to compare with the inequality for the second part derived below.

For the second part, we proceed as follows

$$E\left[(Z + \alpha\sqrt{m})^2 e^{-Z^2}\right] = EZ^2 e^{-Z^2} + 2\alpha\sqrt{m}EZe^{-Z^2} + \alpha^2 m E e^{-Z^2}$$

Since

$$E e^{-Z^2} = \frac{1}{\sqrt{1 + 2\sigma^2}} e^{-\frac{\nu^2}{1+2\sigma^2}}$$

$$EZ e^{-Z^2} = \frac{\nu}{(1 + 2\sigma^2)^{3/2}} e^{-\frac{\nu^2}{1+2\sigma^2}}$$

$$EZ^2 e^{-Z^2} = \frac{1}{(1 + 2\sigma^2)^{3/2}}\left(\sigma^2 + \frac{\nu^2}{1 + 2\sigma^2}\right) e^{-\frac{\nu^2}{1+2\sigma^2}}$$

Thus

$$E\left[(Z + \alpha\sqrt{m})^2 e^{-Z^2}\right]$$

$$= e^{-\alpha^2 m}\left(\frac{m-i}{m+i}\right)^{3/2}\left[\frac{i}{m-i} + \alpha^2 m\left(\frac{m}{m+i} - 2\sqrt{\frac{m}{m-i}} + \frac{m+i}{m-i}\right)\right] e^{\frac{\alpha^2 m}{m+i} i}$$

$$\leq e^{-\alpha^2 m}(m-1)(1 + 2\alpha^2 m)e^{\alpha^2 i}.$$

Consequently,

$$E\left[\Phi(\alpha\sqrt{m}) - \Phi\left(\frac{\alpha m - S_i}{\sqrt{m-i}}\right)\right]^2 \leq \frac{1}{\pi}e^{-\alpha^2 m}(1 + 2\alpha^2 m)e^{\alpha^2 i}.$$

Finally, we use the left inequality from (5.34) to get for large n and $i = 1, \ldots, m-1$

$$E(\tilde{h}_{m,i}^2) \leq \frac{1}{\pi}e^{-\alpha^2 m}(1 + 2\alpha^2 m)e^{\alpha^2 i}\sqrt{2\pi}e^{\frac{\alpha^2 m}{2}}\frac{m^{3/2}\alpha^3}{\alpha^2 m - 1} \leq C\left(e^{\alpha^2}\right)^i.$$

Additionally, for $i = m$ we have

$$E(\tilde{h}_{m,m}^2) = E\left(\frac{I(X_1 + \ldots + X_m > \alpha m)}{1 - \Phi(\alpha\sqrt{m})} - 1\right)^2 = \frac{\Phi(\alpha\sqrt{m})}{1 - \Phi(\alpha\sqrt{m})}.$$

Again using the left inequality from (5.34) we get for large n that

$$E(\tilde{h}_{m,m}^2) < \sqrt{2\pi}e^{\frac{\alpha^2 m}{2}}\frac{\alpha^3 m^{3/2}}{\alpha^2 m - 1} = \sqrt{2\pi}\left[e^{-\frac{\alpha^2 m}{2}}\frac{\alpha^3 m^{3/2}}{\alpha^2 m - 1}\right]e^{\alpha^2 m} \le C\left(e^{\alpha^2}\right)^m.$$

Thus (5.32) is satisfied and we conclude that the assumptions of Theorem 5.2.2 are satisfied. Note that by (5.35)

$$E(h_m)\sqrt{2\pi}\alpha\sqrt{m}e^{\frac{\alpha^2 m}{2}} \to 1.$$

Consequently, if $m/n \to \lambda > 0$ then using Theorem 5.2.2 we obtain

$$\frac{\sqrt{m}e^{\frac{\alpha^2 m}{2}}\mathcal{W}(n,\alpha)}{\binom{n}{m}m!} \xrightarrow{d} \frac{1}{\alpha\sqrt{2\pi}}\exp\left[\sqrt{\lambda}\,\mathcal{Z} - \frac{\lambda(e^{\alpha^2}-1)}{2}\right],$$

where $\mathcal{Z}$ is a zero mean Gaussian variable with variance $Var(\mathcal{Z}) = e^{\alpha^2} - 1$.
Similarly, if $m/n \to 0$ then Theorem 5.2.1 implies

$$\sqrt{\frac{n}{m}}\left(\frac{\mathcal{W}(n,\alpha)}{(1-\Phi(\alpha\sqrt{m}))\binom{n}{m}m!} - 1\right) \xrightarrow{d} \mathcal{Z}.$$

Example 5.3.4 (Number of perfect matchings with few light edges). Consider again $K(m,n)$ and let F denotes the common distribution function of the random weights $X_{i,j}$'s. We are interested in a number $\mathcal{R}_n(r,\alpha)$ of perfect matchings for which the r-th smallest weight (out of m weights) exceeds a given threshold α. That is

$$\mathcal{R}_n(r,\alpha) = Per(h_m)$$

for $h_m(x_1,\ldots,x_m) = I(x_{r:m} > \alpha)$, where $x_{r:m}$ denotes the rth order statistic out of m observations.
Note that

$$I(x_{r:m} > \alpha) \tag{5.36}$$

$$= \sum_{s=0}^{r-1}\sum_{1\le i_1<\ldots<i_s\le m}\prod_{j\in\{i_1,\ldots,i_s\}}I(x_j \le \alpha)\prod_{l\in\{1,\ldots,m\}\setminus\{i_1,\ldots,i_s\}}I(x_l > \alpha),$$

where the summand for $s = 0$ is understood as $\prod_{l=1}^m I(x_l > \alpha)$. Thus, by (5.36) we get

$$E(h_m) = \sum_{s=0}^{r-1}\binom{m}{s}F^s(\alpha)\bar{F}^{m-s}(\alpha),$$

where $\bar{F} = 1 - F$.
In order to calculate $h_{m,i}$ it is helpful to observe that the set

$$\{(x_1,\ldots,x_m) : \; x_{r:m} > \alpha\}$$

can be decomposed in disjoint sets of the form

$$\{(x_1,\ldots,x_m) : \; \text{exactly } s \text{ elements out of } (x_{k+1},\ldots,x_m) \text{ are } \le \alpha \text{ and } x_{r-s:k} > \alpha \},$$

for $s = 0,1,\ldots,r-1$.

We assume here that $x_{j:n} = -\infty$ if $j \le 0$ and $x_{j:n} = \infty$ if $j > n$. Since

$$h_{m,i}(x_1,\ldots,x_i) = E(h_m(x_1,\ldots,x_i,X_{i+1},\ldots,X_m)),$$

where $X_{i+1},\ldots,X_m$ are iid with the df F, then by the above remark it follows that

$$h_{m,i}(x_1,\ldots,x_i) = \sum_{s=0}^{r-1} \binom{m-i}{s} F^s(\alpha) \bar{F}^{m-i-s}(\alpha) I(x_{r-s:i} > \alpha)$$

Thus

$$\tilde{h}_{m,i}(x_1,\ldots,x_i)$$

$$= \frac{\sum_{s=0}^{r-1} \binom{m-i}{s} F^s(\alpha) \bar{F}^{m-i-s}(\alpha) I(x_{r-s:i} > \alpha)}{\sum_{s=0}^{r-1} \binom{m}{s} F^s(\alpha) \bar{F}^{m-s}(\alpha)} - 1$$

$$= \frac{1}{\bar{F}^i(\alpha)} \frac{\sum_{s=0}^{r-1} \binom{m-i}{s} F^s(\alpha) \bar{F}^{r-1-s}(\alpha) I(x_{r-s:i} > \alpha)}{\sum_{s=0}^{r-1} \binom{m}{s} F^s(\alpha) \bar{F}^{r-1-s}(\alpha)} - 1 \to \frac{I(x_{1:i} > \alpha)}{\bar{F}^i(\alpha)} - 1.$$

Hence by (5.9) we get

$$g_{m,k}(x_1,\ldots,x_k) \to \prod_{i=1}^{k} \left(\frac{I(x_i > \alpha)}{\bar{F}(\alpha)} - 1 \right)$$

$$= \left(\frac{F(\alpha)}{\bar{F}(\alpha)} \right)^{k/2} \prod_{i=1}^{k} \phi^{\otimes k}(x_1,\ldots,x_k) = g_k(x_1,\ldots,x_k),$$

where

$$\phi(x) = \frac{I(x > \alpha) - \bar{F}(\alpha)}{\sqrt{F(\alpha)\bar{F}(\alpha)}}$$

is standardized.

Note that $|g_k| \le \left(\max\{1, \frac{F(\alpha)}{\bar{F}(\alpha)}\} \right)^k$ and thus (5.26) is satisfied. Also

$$|\tilde{h}_{m,i}(x_1,\ldots,x_i)| \le \frac{1}{\bar{F}^i(\alpha)} \frac{\sum_{s=0}^{r-1} \binom{m-i}{s} F^s(\alpha) \bar{F}^{r-1-s}(\alpha)}{\sum_{s=0}^{r-1} \binom{m}{s} F^s(\alpha) \bar{F}^{r-1-s}(\alpha)} + 1 < 2 \left(\frac{1}{\bar{F}(\alpha)} \right)^i,$$

and by Remark 5.2.2 it follows that the assumptions of Theorem 5.2.2 are satisfied. Since

$$\frac{E(h_m)}{m^{r-1}F^{r-1}(\alpha)\bar{F}^{m-r+1}(\alpha)} \to 1,$$

in the case $m_n/n \to \lambda$ Theorem 5.2.2 implies

$$\frac{\mathcal{R}_n(r,\alpha)}{\binom{n}{m}m!m^{r-1}\bar{F}^m(\alpha)} \xrightarrow{d} \left(\frac{F(\alpha)}{\bar{F}(\alpha)}\right)^{r-1} e^{\sqrt{\lambda}\mathcal{Z} - \frac{\lambda F(\alpha)}{2\bar{F}(\alpha)}},$$

where $\mathcal{Z}$ is a zero mean Gaussian variable with variance $Var(\mathcal{Z}) = \frac{F(\alpha)}{\bar{F}(\alpha)}$.

Note that the case $r = 1$ is the case of a classical permanent (Definition 1.20) since the kernel h_m is a product of the form $\prod_{i=1}^m I(x_i > \alpha)$. Thus the above asymptotics in this particular case follows from Theorem 3.4.3.

Alternatively, if $m/n \to 0$ then using Theorem 5.2.1 we get

$$\sqrt{\frac{n}{m}}\left(\frac{\mathcal{R}_n(r,\alpha)}{\binom{n}{m}m!\sum_{s=0}^{r-1}\binom{m}{s}F^s(\alpha)\bar{F}^{m-s}(\alpha)} - 1\right) \xrightarrow{d} \mathcal{Z}.$$

Example 5.3.5 (Sum of products of zero-mean weights for perfect matchings).
Let $X_{i,j}$'s have zero mean and unit variance. Let $r \geq 1$ be fixed. For $K(m,n)$ consider $Per(h_m)$ with

$$h_m(x_1,\ldots,x_m) = \sum_{l=r}^m \sum_{1 \leq j_1 < \ldots < j_l \leq m} \prod_{s=1}^l x_{j_s}.$$

Obviously, $E(h_m) = 0$. Moreover, $h_i(x_1,\ldots,x_i) = 0$ for $i = 1,\ldots,r-1$, and

$$h_{m,i}(x_1,\ldots,x_i) = \sum_{l=r}^i \sum_{1 \leq s_1 < \ldots < s_l \leq i} \prod_{w=1}^l x_{s_w}$$

for $i = r, r+1, \ldots, m$, i.e. r is the common degeneracy level of all P-statistics in this example. Note that $h_{m,i}$'s do not depend on m. Using (5.9) we obtain

$$g_{m,k}(x_1,\ldots,x_k) = \sum_{i=r}^k (-1)^{k-i} \sum_{1 \leq j_1 < \ldots < j_i \leq k} \sum_{l=r}^i \sum_{1 \leq s_1 < \ldots < s_l \leq i} \prod_{w=1}^l x_{j_{s_w}}$$

$$= \sum_{i=r}^k \sum_{1 \leq j_1 < \ldots < j_i \leq k} \prod_{w=1}^i x_{j_w} \left(\sum_{l=0}^{k-i} \binom{k-i}{l}(-1)^{k-(i+l)}\right).$$

Noting that the expression in the parantheses in the above formula is zero except in the case $i = k$ we obtain

$$g_{m,k}(x_1,\ldots,x_k) = \prod_{l=1}^k x_l = g_k(x_1,\ldots,x_k),$$

i.e. $\phi(x) = x$ (standardized). Since $E(g_{m,k}^2) = E(g_k^2) = 1$ the assumptions of Theorems 5.2.2 and 5.2.1 apparently are satisfied.

Consequently, for $m/n \to \lambda > 0$,

$$\frac{Per(h_m)}{\binom{n}{m}m!} \xrightarrow{d} e^{\sqrt{\lambda}\mathcal{N}-\frac{\lambda}{2}} - \sum_{k=0}^{r-1} \frac{\lambda^{k/2}}{k!} H_k(\mathcal{N}).$$

In particular, if $r = 2$ the limiting law is the distribution of the random variable $e^{\sqrt{\lambda}\mathcal{N}-\frac{\lambda}{2}} - 1 - \sqrt{\lambda}\mathcal{N}$.

For $m/n \to 0$

$$\sqrt{\frac{n}{m}} \frac{Per(h_m)}{\binom{n}{m}m!} \xrightarrow{d} H_c(\mathcal{N}).$$

Let us note that in the first four examples above we had:

$$\frac{h_{m,i}(x_1,\ldots,x_i)}{E(h_m)} \to \frac{1}{L}\sum_{r=1}^{L}\prod_{l=1}^{i}\psi_r(x_l)$$

for some natural number L, a positive number α and some functions ψ_r, satisfying $E(\psi_r^2(X)) = 1$, $r = 1,\ldots,L$. If this is the case then in general

$$g_{m,k}(x_1,\ldots,x_k) \to \frac{1}{L}\sum_{r=1}^{L}\prod_{l=1}^{k}(\psi_r(x_l) - 1) = g_k(x_1,\ldots,x_k).$$

Consequently, if only the assumptions of Theorem 5.2.2 are satisfied then

$$\frac{Per(h_m)}{\binom{n}{m}m!E(h_m)} \to \frac{1}{L}\sum_{r=1}^{L} e^{\sqrt{\lambda}Z_r - \frac{\lambda(E\psi_r^2(X)-1)}{2}} \tag{5.37}$$

where $(Z_1,\ldots,Z_L)$ is a zero-mean Gaussian vector with covariances

$$E(Z_rZ_s) = E(\psi_r(X)\psi_s(X)) - 1.$$

If in the situation just described the edges $[y_{i,j}]$ in a bipartite graph appear independently with the same probability q and independently of the weight matrix $\mathbb{X}^{(\infty)}$, then we have to modify the kernel of a P-statistic as follows

$$\hat{h}_m((x_1,y_1),\ldots,(x_m,y_m)) = h_m(x_1,\ldots,x_m)\prod_{l=1}^{m}I(y_l = 1),$$

where $y_l = 1$ if the respective edge is present in the graph, otherwise $y_l = 0$ (see also Section 1.7). Then, it is easy to see that

$$\frac{\hat{h}_{m,i}((x_1,y_1),\ldots,(x_i,y_i))}{E(\hat{h}_m)}$$

$$=\frac{h_{m,i}(x_1,\ldots,x_i)}{E(h_m)}\prod_{l=1}^{i}\frac{I(y_l=1)}{q}\to\frac{1}{L}\sum_{r=1}^{L}\prod_{l=1}^{i}\psi_r(x_l)\frac{I(y_l=1)}{q}.$$

Consequently,

$$g_{m,k}((x_1,y_1),\ldots,(x_k,y_k))\to\frac{1}{L}\sum_{r=1}^{L}\prod_{l=1}^{k}\left(\psi_r(x_l)\frac{I(y_l=1)}{q}-1\right)$$

$$=g_k((x_1,y_1),\ldots,(x_k,y_k)).$$

In this situation the convergence result (5.37) has to be modified as follows. If the assumptions of Theorem 5.2.2 are satisfied then

$$\frac{Per(\hat{h}_m)}{\binom{n}{m}m!q^m E(\hat{h}_m)}\to\frac{1}{L}\sum_{r=1}^{L}\exp\left(\sqrt{\lambda}\hat{Z}_r-\frac{\lambda(E\psi_r^2(X)-q)}{2q}\right),\qquad(5.38)$$

where $(\hat{Z}_1,\ldots,\hat{Z}_L)$ is a zero-mean Gaussian random vector with covariances

$$E(\hat{Z}_r\hat{Z}_s)=\frac{E(\psi_r(X)\psi_s(X))}{q}-1.$$

For instance if the edges appear at random, as described above, in the Examples 5.3.2–5.3.4 we obtain the following modifications of the limiting behavior for $m/n\to\lambda>0$:

In Example 5.3.2 for the number of red edges $\mathcal{K}(n,\alpha)$:

$$\frac{\mathcal{K}(n,\alpha)}{\binom{n}{m}m!\binom{m}{L_n}p^{L_n}(1-p)^{m-L_n}q^m}\to\exp\left(\sqrt{\lambda}\,\mathcal{Z}-\frac{\lambda[(\alpha-p)^2+(1-q)p(1-p)]}{2qp(1-p)}\right),$$

where $\mathcal{Z}$ is a zero mean Gaussian variable with variance $Var(\mathcal{Z})=\frac{(\alpha-p)^2}{qp(1-p)}+\frac{1-q}{q}$.

In Example 5.3.3 for the number of heavy normal matchings $\mathcal{W}(n,\alpha)$:

$$\frac{\sqrt{m}e^{\frac{\alpha^2 m}{2}}\mathcal{W}(n,\alpha)}{\binom{n}{m}m!q^m}\xrightarrow{d}\frac{1}{\alpha\sqrt{2\pi}}\exp\left[\sqrt{\lambda}\mathcal{Z}-\frac{\lambda(e^{\alpha^2}-q)}{2q}\right],$$

where $\mathcal{Z}$ is a zero mean Gaussian variable with variance $Var(\mathcal{Z})=\frac{e^{\alpha^2}-q}{q}$.

Example 5.3.4 for the number of matchings with few light edges $\mathcal{R}_n(r,\alpha)$:

$$\frac{\mathcal{R}_n(r,\alpha)}{\binom{n}{m}m!m^{r-1}[q\bar{F}(\alpha)]^m}\xrightarrow{d}\left(\frac{F(\alpha)}{\bar{F}(\alpha)}\right)^{r-1}e^{\sqrt{\lambda}\mathcal{Z}-\frac{\lambda(1-q\bar{F}(\alpha))}{2q\bar{F}(\alpha)}},$$

where $\mathcal{Z}$ is a zero mean Gaussian variable with the variance $Var(\mathcal{Z})=\frac{1-q\bar{F}(\alpha)}{q\bar{F}(\alpha)}$.

5.4 Bibliographic Details

Multiple Wiener-Itô integral was introduced by Itô (1951) who considerably modified the original idea of homogeneous chaos introduced by Wiener (1938). A detailed description can be found in the monograph by Major (1981) which is devoted solely to this issue. A concise and quite accessible exposition of the topic has been given recently in Kuo (2006). Here we adopted an approach which borrowed the first step from Dynkin and Mandelbaum (1983) and was especially convenient in the context of symmetric functions, which were of the primary interest.

The fundamental result on the weak convergence of non-degenerate U-statistics was obtained by Hoeffding (1948) in his seminal paper on the subject. Subsequent results for degenerate cases were obtained by Serfling (1980), Rubin and Vitale (1980) and Dynkin and Mandelbaum (1983) who were the first to introduce the idea of writing a limiting distribution in terms of multiple Wiener-Itô integrals. For a general review of the subject of multiple stochastic integration in the context of symmetric functions see e.g., monographs by Major (1981) or Lee (1990, chapter 4). There are also several other good monographs that consider, among other topics, the issue of weak convergence of U-statistics both in the case of finite order kernels as well as the sequences of kernels of increasing dimensions, like e.g., Koroljuk and Borovskich (1994). Some interesting results related to the limiting behavior of so called 'decoupled' U-statistics, which are similar to our P-statistics in the case of matrix $\mathbb{X}^{(\infty)}$ having independent and identically distributed entries, are presented in de la Peña and Giné (1999, chapter 4). For some related results on the weak convergence of both classical and decoupled U-statistics the reader may also wish to consult e.g., the papers of Giné and Zinn (1994), Latała and Zinn (2000), and Giné et al. (2001).

The material on weak convergence of P-statistics presented in this chapter is essentially new and comes from the yet unpublished manuscript Rempała and Wesołowski (2007).

6

Permanent Designs and Related Topics

6.1 Incomplete U-statistics

In the current chapter we shall explore a different view of a P-statistic, namely as a certain type of an *incomplete* U-statistic a notion that we now proceed to describe. Recall that in the previous chapters we have defined a U-statistic $U_l^{(k)}$ as

$$U_l^{(k)} = \binom{l}{k}^{-1} \pi_k^l(h)$$

for any symmetric kernel h. For given integers $1 \le k \le l$ let $\mathcal{S}_{l,k}$ denote as before a set of all k-subsets of $\{1, \ldots, l\}$, i.e., a collection of all sets $S = \{i_1, \ldots, i_k\}$ of distinct indices selected from $\{1 \ldots, l\}$. In what follows we shall also often write $h(S)$ for $h(Y_{i_1} \ldots, Y_{i_k})$. In the above notation we have

$$U_l^{(k)} = \binom{l}{k}^{-1} \sum_{S \in \mathcal{S}_{l,k}} h(S).$$

Even though U-statistics are relatively simple and thus theoretically appealing probabilistic objects their practical use is severely limited due to the fact that for large values of k and l the number of needed averagings $\binom{l}{k}$ may be very large and the actual evaluation of a U-statistic can be quite onerous especially when considering infinite order U-statistics, like e.g., elementary symmetric polynomial of increasing order discussed in detail in Chapter 3, where the dimension k of a kernel function is growing with the sample size l. In order to address the problem of these computational difficulties Blom (1976) has suggested considering "incomplete" U-statistics, where the kernel function is averaged over only some appropriately chosen small subset (a design) of all $\binom{l}{k}$ averagings. The idea of Blom has turned out to be closely related to the general statistical theory of experimental designs and has lead over the next decade to a rapid development of the theory of incomplete U-statistics based on a variety of designs.

The main drawback of the theory of incomplete U-statistics seems to be in a lack of a sufficiently general and yet relatively simple technique of generating incomplete designs (except perhaps for the random selection) that would work well for a very wide class of U-statistics, including the case when $k \to \infty$. In this chapter we shall present a class of designs which in some circumstances may work well for a large class of kernels.

If we assume that $E\, h_k^2 < \infty$ for $l \geq 1$, then the variance of a U-statistic of degree $r - 1$ is given by the formula (2.14) of Chapter 2.

$$Var\, U_l^{(k)} = \sum_{c=r}^{k} \binom{l}{c}^{-1} \binom{k}{c}^2 Var\, g_c. \qquad (6.1)$$

A convenient alternative form is obtained by rewriting the above in terms of the variances of the conditional kernels h_c, $c = 1, \ldots, k$, as follows

$$Var\, U_l^{(k)} = \binom{l}{k}^{-1} \sum_{c=r}^{k} \binom{k}{c} \binom{l-k}{k-c} Var\, h_c. \qquad (6.2)$$

To see this, recall from Chapter 1 that

$$Var\, g_c = \sum_{\nu=1}^{c} (-1)^{c-\nu} \binom{c}{\nu} Var\, h_\nu. \qquad (6.3)$$

Now, substituting (6.3) into (6.1) gives after several lines of algebra (6.2).

Note that for $r = 1$, that is for non-degenerate U-statistics, the relation (6.2) along with the property (1.18) of the sequence $Var\, h_\nu$ $\nu = 1, \ldots, k$ and the hypergeometric summation formula yields the bounds

$$\frac{k^2}{l} Var\, h_1 \leq Var\, U_l^{(k)} \leq \frac{k}{l} Var\, h_k. \qquad (6.4)$$

The following definition shall be essential in the further discussions in this chapter.

Definition 6.1.1 (Incomplete U-statistic). *For any class of k-subsets $\mathcal{D} \subseteq \mathcal{S}_{l,k}$ an incomplete U-statistic is defined as*

$$U_{\mathcal{D}}^{(k)} = \frac{1}{|\mathcal{D}|} \sum_{S \in \mathcal{D}} h(S) \qquad (6.5)$$

where $|\mathcal{D}|$ denotes the number of elements in $\mathcal{D}$.

The set $\mathcal{D}$ is often referred to as a *design* of an incomplete U-statistic. The variance formulae (6.1) and (6.2) have their analogues for $U_{\mathcal{D}}^{(k)}$. We state and prove this for the record as follows.

Theorem 6.1.1. *Let f_c for $1 \le c \le k$ be the number of pairs S_1, S_2 in the design $\mathcal{D}$ having exactly c elements in common. Then*

$$Var\, U_{\mathcal{D}}^{(k)} = \frac{1}{|\mathcal{D}|^2} \sum_{c=r}^{k} f_c Var\, h_c. \tag{6.6}$$

Additionally, for $S \in \mathcal{S}_{l,\nu}$, where $1 \le \nu \le k$, we define $n(S)$ as the number of k-subsets in the design $\mathcal{D}$ which contain S, i.e. $n(S) = \#\{S' \in \mathcal{D} : S \subset S'\}$.
  Then

$$Var\, U_{\mathcal{D}}^{(k)} = \frac{1}{|\mathcal{D}|^2} \sum_{\nu=r}^{k} B_\nu Var\, g_\nu \tag{6.7}$$

where

$$B_\nu = \sum_{S \in \mathcal{S}_{l,\nu}} n^2(S) = \sum_{c=\nu}^{k} f_c \binom{c}{\nu}. \tag{6.8}$$

Moreover, for any $\nu = 1 \ldots, k$

$$A_\nu = \sum_{S \in \mathcal{S}_{l,\nu}} n(S) = |\mathcal{D}| \binom{k}{\nu}. \tag{6.9}$$

Proof. Denote $|\mathcal{D}| = d$. First we shall argue (6.6). To this end note that for any two sets $U, T \in \mathcal{S}_{l,k}$ with exactly c elements in common we have $Cov\,(h(U), h(T)) = Var\, h_c$. Therefore

$$Var\, U_{\mathcal{D}}^{(k)} = d^{-2} \sum_{U \in \mathcal{D}} \sum_{T \in \mathcal{D}} Cov\,(h(U), h(T)) = d^{-2} \sum_{c=1}^{k} f_c Var\, h_c \tag{6.10}$$

which establishes (6.6).

In order to argue (6.7) we define $B_\nu = \sum_{S \in \mathcal{S}_{l,\nu}} n^2(S)$ and consider the design incidence matrix, given by $N = [n_{ij}]_{l \times d}$ where

$$n_{ij} = \begin{cases} 1 & \text{if index } i \text{ is in set } j, \\ 0 & \text{otherwise.} \end{cases} \tag{6.11}$$

Note that the (j_1, j_2)th entry of $N^T N$ is $\sum_i n_{ij_1} n_{ij_2} = |S_{j_1} \cap S_{j_2}|$, where $S_{j_1}, S_{j_2} \in \mathcal{D}$. Thus f_c is the number of elements in the matrix $N^T N$ equal to c. In order to find the formula for f_c we shall use the Stirling numbers. Let $s_{\mu,\nu}^{(1)}$ and $s_{\nu,\mu}^{(2)}$ be the Stirling numbers of the first and second kind, respectively. Recall that they are defined by the equations

$$x^\nu = \sum_{\mu=1}^{\nu} s_{\nu,\mu}^{(2)} x(x-1) \cdots (x - \mu + 1)$$

and

$$x(x-1)\ldots(x-\mu+1) = \sum_{\nu=1}^{\mu} s_{\mu,\nu}^{(1)} x^{\nu}$$

and satisfy the relationship (see, e.g. Abramowitz and Stegun 1964).

$$\sum_{\nu=\mu}^{\gamma} s_{\nu,\mu}^{(2)} s_{\gamma,\nu}^{(1)} = \begin{cases} 1 & \gamma = \mu, \\ 0 & \gamma \neq \mu. \end{cases} \tag{6.12}$$

Additionally, recall also that $s_{\nu,\mu}^{(2)}$ gives the total number of different ways of partitioning a set of size ν into μ non-empty sets.

In view of the above interpretation of f_c in terms of the elements of matrix $N^T N$, we may write for any $\nu \geq 1$

$$\sum_{c=1}^{k} c^{\nu} f_c = \sum_{j_1=1}^{d} \sum_{j_1=1}^{d} \left(\sum_{i=1}^{l} n_{ij_1} n_{ij_2} \right)^{\nu}$$

$$= \sum_{i_1=1}^{l} \cdots \sum_{i_\nu=1}^{l} \left(\sum_{j=1}^{d} n_{i_1 j} \cdots n_{i_\nu j} \right)^{2}$$

$$= \sum_{\mu=1}^{\nu} \mu! s_{\nu,\mu}^{(2)} B_\mu \tag{6.13}$$

where in the last equality we have used the interpretation of the Stirling numbers in terms of the numbers of partitions, as well as the fact that the entries n_{ij} are binary. Now we multiple both sides of (6.13) by $s_{\gamma,\nu}^{(1)}$ and sum over ν from one up to γ. This, in view of (6.12), yields for $\gamma = 1, \ldots, k$

$$B_\gamma = \sum_{c=\gamma}^{k} f_c \binom{c}{\gamma} \tag{6.14}$$

and thus also (6.8). Multiplying (6.14) on both sides by $(-1)^{\gamma-\nu}\binom{\gamma}{\nu}$, summing over γ from ν up to k and using the identity

$$\sum_{\gamma=\nu}^{c} (-1)^{\gamma-\nu} \binom{\gamma}{\nu} \binom{c}{\gamma} = \begin{cases} \binom{c}{\nu} & \text{if } c = \nu, \\ 0 & \text{otherwise,} \end{cases}$$

gives for $c = 1, \ldots, k$

$$f_c = \sum_{\gamma=c}^{k} (-1)^{\gamma-c} \binom{\gamma}{c} B_\gamma.$$

In view of the above, as well as (6.10)

$$Var\, U_{\mathcal{D}}^{(k)} = d^{-2} \sum_{c=1}^{k} f_c Var\, h_c$$

$$= d^{-2} \sum_{\gamma=1}^{k} \left(\sum_{c=1}^{\gamma} (-1)^{\gamma-c} \binom{\gamma}{c} Var\, h_c \right) B_\gamma$$

$$= \sum_{\gamma=1}^{k} B_\gamma Var\, g_c$$

where the last equality follows from (6.3). The formula (6.7) is thus proved. It remains to show (6.9). Note that for $\nu = 1, \ldots, k$ we have

$$dk^\nu = \sum_{j=1}^{d} \left(\sum_{i=1}^{l} n_{ij} \right)^\nu$$

$$= \sum_{i_1=1}^{n} \cdots \sum_{i_\nu=1}^{n} \left(\sum_{j=1}^{d} n_{i_1 j} \cdots n_{i_\nu j} \right)$$

$$= \sum_{\mu=1}^{\nu} \mu! s_{\nu,\mu}^{(2)} A_\mu$$

which leads to (6.9) in exactly the same way in which (6.13) lead to (6.14). □

Comparing (6.1) and (6.6) or (6.2) and (6.7) it is not difficult to see that for $\mathcal{D} \subseteq \mathcal{S}_{l,k}$ we must always have $Var\, U_l^{(k)} \leq Var\, U_{\mathcal{D}}^{(k)}$ with equality only if $\mathcal{D} = \mathcal{S}_{l,k}$. Therefore, the incomplete U-statistic is always less efficient than the complete one. However, for appropriately chosen design $\mathcal{D}$, the increase in the variance of U may be not too large and the loss of the estimation precision may be offset by the considerable simplification of the statistic. The problem of an appropriate choice of $\mathcal{D}$ is therefore central to the theory of incomplete U-statistics (cf. e.g., Lee 1982).

6.2 Permanent Design

It turns out that one possible solution to the problem of generating simple designs is via a concept of a generalized permanent function and a P-statistic. Since the design is cast in the setting of selecting subsets from a matrix of mn independent and identically distributed random variables, hence in this chapter we again strengthen the exchangeability assumptions (A1-A2) to the general assumption that all the entries of $\mathbb{X}^{(\infty)}$ are independent. The "permanent" design for selecting elements from $\mathbb{X}^{(m,n)}$ into the averaging scheme turns out to be quite reasonable (for instance, "balanced" in the experimental design sense) due to the inherited symmetry. In particular, under some regularity conditions imposed on the kernel functions of the corresponding U-statistics it turns out to be always asymptotically efficient, in the usual sense of the asymptotic relative efficiency (ARE), when compared with a full design of all $\binom{mn}{m}$ subsets. It is remarkable that this efficiency property of the design holds even in the case when both $m, n \to \infty$. In some cases the permanent design can also be shown to be more effective than any other design of the same size, including a random one.

Let us consider a U-statistic of degree m based on a double-indexed sequence (matrix) of iid real random variables $\mathbb{X}^{(m,n)} = [X_{i,j}]$ with $i = 1, \ldots, m; j = 1 \ldots, n$ and $1 \leq m \leq n$

$$U_{mn}^{(m)} = \binom{mn}{m}^{-1} \sum_{S \in \mathcal{S}_{mn,m}} h(S) \tag{6.15}$$

where $\mathcal{S}_{mn,m}$ is a set of all m-subsets of the ordered pairs of indices $\{(i,j)|1 \leq i \leq m; 1 \leq j \leq n\}$. Note that the above definition is equivalent to that of a U-statistic $U_l^{(k)}$ with $k = m$, $l = mn$ and the sequence $X_{1,1}, \ldots, X_{1,n}, X_{2,1}, \ldots, X_{2,n}, \ldots, X_{m,1} \ldots, X_{m,n}$ obtained by vectorization of the matrix $[X_{i,j}]$.

Any design $\mathcal{D} \subseteq \mathcal{S}_{mn,m}$ we shall call *rectangular*. Due to the symmetry assumption about the kernel function it is reasonable to restrict attention to *equireplicate rectangular designs*, i.e., such that $\forall_{(i,j)}$ the number $n(i,j)$ of subsets in $\mathcal{D}$ containing (i,j) is constant and equal s. Note that such designs must satisfy:

$$smn = dm \tag{6.16}$$

where $d = |\mathcal{D}|$. The above follows by representing in two ways a sum of multiplicities of all singletons in the design $\mathcal{D}$.

The following definitions will be useful in the sequel.

Definition 6.2.1 (Meager design). *We will say that a design $\mathcal{D} \subseteq \mathcal{S}_{mn,m}$ is meager if for any $\nu \in \{1, \ldots, m\}$ the number of elements in $\mathcal{S}_{mn,\nu}$ being subsets of sets belonging to $\mathcal{D}$ equals $\binom{n}{\nu}\binom{m}{\nu}\nu!$*

In particular, for any meager design $\mathcal{D}$ we have $|\mathcal{D}| = n^{[m]} = \binom{n}{m}m!$. Let us also note that the notion of a meager design in our setting is related to that of a *strongly regular graph design* (see Subsection 6.5.1) known in the experimental design theory.

Definition 6.2.2 (Permanent scheme design). *A design $\mathcal{D} \subseteq \mathcal{S}_{mn,m}$ is a permanent scheme if it consists of all m-subsets of the form*

$$\{(1, i_1), (2, i_2), (3, i_3) \ldots, (m, i_m)\} \tag{6.17}$$

where $\{i_1, i_2, \ldots, i_m\} \subset \{1, 2, \ldots, n\}$.

In the sequel we refer to any m-set of the form (6.17) as m-*transversal* of an $m \times n$ matrix. Thus we may say that permanent scheme design is simply a collection of all m-transversals (of an $m \times n$ matrix).

Below we shall denote a permanent scheme design by $\mathcal{D}_{per}$ and refer to it for brevity as *permanent design*. Accordingly, throughout this chapter we shall also refer to any associated P-statistic as (incomplete) U-statistic of permanent design (USPD) and denote it by $U_{\mathcal{D}_{per}}^{(m)}$.

We note that according to the earlier definition USPD is simply a P-statistic based on a matrix of all iid entries.

The following theorem describes some of the basic properties of a permanent design.

Theorem 6.2.1. *The design $\mathcal{D}_{per}$ is balanced (equireplicate) and meager. It is also a minimum variance design in the class of meager designs, that is, for fixed integers $1 \leq m \leq n$ it minimizes the variance of an incomplete analogue of (6.15) among all meager designs.*

Proof. First, we show that $\mathcal{D}_{per}$ is meager. That is, for given ν $(1 \leq \nu \leq m)$ we show that the number of different sets $S \in \mathcal{S}_{mn,\nu}$ being subsets of sets belonging to $\mathcal{D}_{per}$ is $\binom{m}{\nu}\binom{n}{\nu}\nu!$. To see this, consider a set of ordered pairs of indices (coordinates) $\{(i,j)|1 \leq i \leq m; 1 \leq j \leq n\}$. Taking any ν-subset is equivalent to fixing the values of ν first coordinates and ν second coordinates, which can be done in $\binom{m}{\nu}\binom{n}{\nu}$ different ways; for the selected ν pairs of indices there is $\nu!$ ways to be included into one of the sets of the form (6.17).

Second, we show that $\mathcal{D}_{per}$ is of minimum variance among meager designs. For a given $S \in \mathcal{S}_{mn,\nu}$ let us consider $n(S)$, i.e. the number of m-subsets in the design $\mathcal{D}_{per}$ which contain S. (cf. Theorem 6.1.1). If S is a subset of one of the m-sets of the form (6.17) then, from the definition of the permanent design it follows that for such S

$$n(S) = \binom{n-\nu}{m-\nu}(m-\nu)! \tag{6.18}$$

On the other hand, if S is not a subset of one of the m-sets of the form (6.17) then $n(S) = 0$. From the relation (6.9) of Theorem 6.1.1 we know that for any design $\mathcal{D} \subseteq \mathcal{S}_{mn,m}$ such that $|\mathcal{D}| = \binom{n}{m}m!$ we have

$$\sum_{S \in \mathcal{S}_{mn,\nu}} n(S) = \binom{m}{\nu}\binom{n}{m}m!.$$

Since the quadratic $\sum x_i^2$ subject to $\sum x_i = c > 0$ is minimized by taking all the x_i's equal, it follows (see (6.8)) that for the class of all meager designs $\mathcal{D}_{per}$ minimizes the quantities B_ν in the formula (6.7) and hence it minimizes also the variance. The fact that $\mathcal{D}_{per}$ is balanced follows directly from (6.18) with $\nu = 1$.   □

It is not difficult to show that there exist designs which are meager, balanced, and have strictly larger variance than the permanent design. Therefore, the above theorem is meaningful.

Example 6.2.1 (Meager non-permanent designs). Let $\mathcal{D} \subseteq S_{2n,2}$ be any balanced design such that $|\mathcal{D}| = n(n-1)$. Then $\mathcal{D}$ is a meager design.

A connection between a USPD and a U-statistic is given by the following

Theorem 6.2.2. *For given fixed integers $1 \leq m \leq n$ let Π be the set of all possible permutations of the elements of $m \times n$ matrix $[X_{i,j}]$ and for given $\sigma \in \Pi$ let $U_{\mathcal{D}_{per}}^{(m)}(\sigma)$ denote a corresponding incomplete U-statistic. Then*

$$U_{mn}^{(m)} = \frac{1}{(mn)!} \sum_{\sigma \in \Pi} U_{\mathcal{D}_{per}}^{(m)}(\sigma).$$

Proof. Direct calculation.   □

The next example illustrates the use of Theorem 6.1.1 in the particular case of a permanent design. This gives an alternative way of arriving at the formula (2.13).

Example 6.2.2 (Variance of USPD). From the general result on the variance of P-statistics under (A1-A2) we know that the variance of a USPD is given by (cf. (2.13))

$$Var\, U_{\mathcal{D}_{per}}^{(m)} = \sum_{\nu=r}^{m} \frac{\binom{m}{\nu}}{\binom{n}{\nu}} \frac{1}{\nu!} Var\, g_{\nu}. \tag{6.19}$$

Note that this formula may also be obtained directly from the general results on the incomplete U-statistics discussed so far. Indeed, in the proof of Theorem 6.2.1 we have shown that $\mathcal{D}_{per}$ is meager and that for any ν ($1 \leq \nu \leq m$) and any $S \in \mathcal{S}_{mn,\nu}$ being a subset of one of the m-sets belonging to $\mathcal{D}_{per}$, $n(S)$ is given by (6.18). This, along with the formula (6.7) entails (6.19), since now

$$B_{\nu} = \binom{m}{\nu} \binom{n}{\nu} \nu! \binom{n-\nu}{m-\nu}^{2} (m-\nu)!^{2}$$

for $1 \leq \nu \leq m$.

As in the case of U-statistics, for a USPD there also exists an alternative formula for its variance in terms of the variances of the appropriate conditional expectations of the kernel function.

$$Var\, U_{\mathcal{D}_{per}}^{(m)} = \frac{1}{\binom{n}{m} m!} \sum_{\nu=r}^{m} \binom{m}{\nu} \Psi(m-\nu, n-\nu) Var\, h_{\nu} \tag{6.20}$$

where

$$\Psi(i,j) = \sum_{c=0}^{i}(-1)^c \binom{i}{c}\binom{j-c}{i-c}(i-c)!$$

for $0 \leq i \leq m-1$; $n-m \leq j \leq n-1$. The formula (6.20) may be obtained from (6.19) with the help of the relation (6.3). Equivalently, it may be inferred directly from (6.6) upon verifying that $f_c = \binom{n}{m}m!\binom{m}{c}\Psi(m-c, n-c)$.

6.3 Asymptotic properties of USPD

The next result shows that asymptotically the variances of USPD and a corresponding complete U-statistic coincide for a large class of kernel functions.

For any incomplete U-statistic (6.5) let ARE be its asymptotic relative efficiency as compared with the complete statistic $U_l^{(k)}$, that is,

$$ARE = \lim_{l \to \infty} \frac{Var\, U_l^{(k)}}{Var\, U_D^{(k)}}.$$

Theorem 6.3.1. *Suppose that for the U-statistic (6.15) of a fixed (i.e., independent of n, m) level of degeneration $r-1$ we have $E\,(h^{(m)})^2 < \infty$, $0 < \liminf Var\, g_r^{(m)}$, and $Var\, g_\nu^{(m)} \leq c_\nu$ for $\nu \geq r$ and some constants c_ν (independent of m, n) which satisfy*

$$\sum_{\nu=r}^{\infty} \frac{c_\nu}{\nu!} < \infty. \tag{6.21}$$

If $n \geq m \to \infty$ and $\frac{m}{n} \to \lambda \geq 0$, then ARE of USPD vis à vis the complete U-statistic (6.15) equals one.

Furthermore, for $r = 1$ the result remains valid if $m = m_1 \geq 1$ is a fixed integer and $n \to \infty$.

Proof. Let us note that by (6.1) and (6.19) we have

$$\frac{Var\, U_{D_{per}}^{(m)}}{Var\, U_{mn}^{(m)}} - 1 = \frac{\sum_{\nu=r}^{m}\left[\frac{\binom{m}{\nu}}{\binom{n}{\nu}}\frac{1}{\nu!} - \frac{\binom{m}{\nu}^2}{\binom{mn}{\nu}}\right]Var\, g_\nu^{(m)}}{\sum_{\nu=r}^{m}\frac{\binom{m}{\nu}^2}{\binom{mn}{\nu}}Var\, g_\nu^{(m)}} \geq 0 \tag{6.22}$$

as each term in the sum in the numerator is non-negative, in view of the inequality $\binom{mn}{\nu} \geq \binom{m}{\nu}\binom{n}{\nu}(\nu!)^{-1}$ which is valid for all integers $1 \leq \nu \leq m \leq n$.

On the other hand, for the expression on the right-hand side of the equality sign in (6.22), we clearly have for sufficiently large n

$$\frac{\sum_{\nu=r}^{m}\left[\frac{\binom{m}{\nu}}{\binom{n}{\nu}}\frac{1}{\nu!}-\frac{\binom{m}{\nu}^2}{\binom{mn}{\nu}}\right]Var\, g_\nu^{(m)}}{\sum_{\nu=r}^{m}\frac{\binom{m}{\nu}^2}{\binom{mn}{\nu}}Var\, g_\nu^{(m)}} \le \frac{\sum_{\nu=r}^{m}\left[\frac{\binom{m}{\nu}}{\binom{n}{\nu}}\frac{1}{\nu!}-\frac{\binom{m}{\nu}^2}{\binom{mn}{\nu}}\right]Var\, g_\nu^{(m)}}{\frac{\binom{m}{r}^2}{\binom{mn}{r}}Var\, g_r^{(m)}}$$

$$\le C\,\frac{\sum_{\nu=r}^{m}\left[\frac{\binom{m}{\nu}}{\binom{n}{\nu}}\frac{1}{\nu!}-\frac{\binom{m}{\nu}^2}{\binom{mn}{\nu}}\right]Var\, g_\nu^{(m)}}{\frac{\binom{m}{r}^2}{\binom{mn}{r}}}$$

$$(6.23)$$

where $C > 0$ is a constant which does not depend on n and m. Let us show that under the assumptions of the theorem the latest expression tends to zero. We will prove this separately for each of the following cases.

Case $\lambda > 0$

Let $\varepsilon > 0$ be arbitrarily small and let m_0 be an integer such that $\sum_{\nu=m_0}^{\infty}(\nu!)^{-1}c_\nu < \varepsilon$ (the existence of m_0 is guaranteed by (6.21)). Since for any fixed integer ν with $r \le \nu \le m_0$ we have that $\binom{m}{\nu}^2/\binom{mn}{\nu} \to \lambda^\nu/\nu!$ and $\frac{\binom{m}{\nu}}{\binom{n}{\nu}} \to \lambda^\nu$ as $n \ge m \to \infty$ with $\frac{m}{n} \to \lambda$, then, for sufficiently large m,n

$$\frac{\sum_{\nu=r}^{m}\left[\frac{\binom{m}{\nu}}{\binom{n}{\nu}}\frac{1}{\nu!}-\frac{\binom{m}{\nu}^2}{\binom{mn}{\nu}}\right]Var\, g_\nu^{(m)}}{\frac{\binom{m}{r}^2}{\binom{mn}{r}}} \le \frac{2r!}{\lambda^r}\sum_{\nu=r}^{m_0}\left[\frac{\binom{m}{\nu}}{\binom{n}{\nu}}\frac{1}{\nu!}-\frac{\binom{m}{\nu}^2}{\binom{mn}{\nu}}\right]c_\nu + \varepsilon\frac{2r!}{\lambda^r}$$

$$\le \varepsilon\frac{2r!}{\lambda^r} + \varepsilon\frac{2r!}{\lambda^r} = \varepsilon\frac{4r!}{\lambda^r}$$

and the assertion follows in view of the relation (6.21) and the arbitrary choice of ε.

Case $\lambda = 0$, $m \to \infty$

In this case,

$$\frac{\sum_{\nu=r}^{m}\left[\frac{\binom{m}{\nu}}{\binom{n}{\nu}}\frac{1}{\nu!}-\frac{\binom{m}{\nu}^2}{\binom{mn}{\nu}}\right]Var\, g_\nu^{(m)}}{\frac{\binom{m}{r}^2}{\binom{mn}{r}}} =$$

$$= \left[\frac{\binom{m}{r}}{\binom{n}{r}}\frac{\binom{mn}{r}}{\binom{m}{r}^2}\frac{1}{r!}-1\right]Var\, g_r^{(m)} \qquad (6.24a)$$

$$+ \sum_{\nu=r+1}^{m}\left[\frac{\binom{m}{\nu}}{\binom{n}{\nu}}\frac{1}{\nu!}\frac{\binom{mn}{r}}{\binom{m}{r}^2}-\frac{\binom{m}{\nu}^2}{\binom{mn}{\nu}}\frac{\binom{mn}{r}}{\binom{m}{r}^2}\right]Var\, g_\nu^{(m)}. \qquad (6.24b)$$

Note that

$$\frac{\binom{m}{r}}{\binom{n}{r}} \frac{\binom{mn}{r}}{\binom{m}{r}^2} \frac{1}{r!} = \frac{\binom{mn}{r}}{r! \binom{n}{r}\binom{m}{r}}$$

$$= \frac{mn(mn-1)\cdots(mn-r+1)}{m(m-1)\cdots(m-r+1)\,n(n-1)\cdots(n-r+1)} \qquad (6.25)$$

$$= \frac{(1-1/mn)\cdots(1-(r-1)/mn)}{(1-1/m)\cdots(1-(r-1)/m)\,(1-1/n)\cdots(1-(r-1)/n)} \to 1$$

as $m, n \to \infty$. Thus, (6.24a) $\to 0$ as $m, n \to \infty$ and we only need to argue that so does (6.24b). To this end, let us note that by (6.25), for sufficiently large m, n,

$$\sum_{\nu=r+1}^{m} \left[\frac{\binom{m}{\nu}}{\binom{n}{\nu}} \frac{1}{\nu!} \frac{\binom{mn}{r}}{\binom{m}{r}^2} - \frac{\binom{m}{\nu}^2}{\binom{mn}{\nu}} \frac{\binom{mn}{r}}{\binom{m}{r}^2} \right] Var\, g_\nu^{(m)}$$

$$\leq \sum_{\nu=r+1}^{m} \frac{\binom{m}{\nu}}{\binom{n}{\nu}} \frac{1}{\nu!} \frac{\binom{mn}{r}}{\binom{m}{r}^2} c_\nu \leq 2 \sum_{\nu=r+1}^{m} \frac{\binom{m}{\nu}}{\binom{n}{\nu}} \frac{r!}{\nu!} \frac{\binom{n}{r}}{\binom{m}{r}} c_\nu$$

$$= 2 \sum_{\nu=r+1}^{m} \frac{\binom{m-r}{\nu-r}}{\binom{n-r}{\nu-r}} \frac{r!}{\nu!} c_\nu \leq 2r! \frac{m-r}{n-r} \sum_{\nu=r+1}^{m} \frac{1}{\nu!} c_\nu \to 0,$$

in view of (6.21) and $\frac{m}{n} \to 0$. The latter inequality follows since for $1 \leq \nu \leq m \leq n$ the expression $\binom{m}{\nu}/\binom{n}{\nu}$ is a non-increasing function of ν. The assertion follows now via (6.22).

Case $r = 1$ and $m = m_1 \geq 1$ is a constant.
If $m_1 = 1$ the result is obvious. If $m_1 > 1$ then proceeding similarly as above we see that for $r = 1$ the expression (6.24a) equals zero and we only need to show that (6.24b) tends to zero as n increases. This is obvious upon noticing that for $\nu = 2, \ldots, m_1$ the expressions in the square brackets are of order $O(n^{-\nu+1})$. □

The above result appears to be quite useful, since, as we have already noted, the size of any meager design $\mathcal{D}$ is $|\mathcal{D}| = \binom{n}{m}m!$ which even for fixed integers $1 \leq m \leq n$ is usually a much smaller number than $\binom{mn}{m}$ – the size of a complete design in (6.15). Asymptotically, this is even more apparent, since by virtue of the Stirling formula we have, for a universal constant $C > 0$,

$$\frac{\binom{n}{m}m!}{\binom{mn}{m}} \leq C\sqrt{m}\, \exp(-m) \to 0 \qquad \text{as} \quad m \to \infty$$

Since for any two sequences (a_k) and (b_k) of nonnegative real numbers satisfying $a_k = \sum_{c=1}^{k} \binom{k}{c} b_c$ for $k \geq 1$, the condition $\sum_{k=1}^{\infty} b_k/k! < \infty$ is equivalent to $\sum_{k=1}^{\infty} a_k/k! < \infty$, in view of the relation (1.19) it is easily seen that a small modification of the above proof allows us for an alternative formulation of Theorem 6.3.1 where the $Var\, g_\nu^{(m)}$'s are replaced with the $Var\, h_\nu^{(m)}$'s. In this alternative form the assumptions of the theorem are perhaps easier to verify and, in particular, it is readily seen that they are satisfied whenever $\liminf_n Var\, h_r^{(m)} > 0$ and $\sup_n E\, (h^{(m)})^2 < \infty$.

As an immediate consequence of Theorem 6.3.1 we obtain the following.

Corollary 6.3.1. *Suppose that the non-degenerate* $(r = 1)$ *U-statistic satisfies the assumptions of Theorem 6.3.1.*
(i) If $n - m \to \infty$ with $\frac{m}{n} \to 0$, then

$$\frac{U_{\mathcal{D}_{per}}^{(m)}}{\left(Var\, U_{\mathcal{D}_{per}}^{(m)}\right)^{1/2}} - \frac{U_{mn}^{(m)}}{\left(Var\, U_{mn}^{(m)}\right)^{1/2}} \xrightarrow{P} 0.$$

(ii) If $n, m \to \infty$ with $\frac{m}{n} \to \lambda > 0$, then

$$U_{\mathcal{D}_{per}}^{(m)} - U_{mn}^{(m)} \xrightarrow{P} 0.$$

Proof. Let us note that for any U-statistic $U_l^{(k)}$ and its incomplete version $U_{\mathcal{D}}^{(k)}$ given by (6.5), due to symmetries involved, we have

$$Cov\,(U_l^{(k)}, U_{\mathcal{D}}^{(k)}) = Cov\,(U_l^{(k)}, h(S)) \qquad \text{for any } S \in \mathcal{D}$$

which entails

$$Cov\,(U_l^{(k)}, U_{\mathcal{D}}^{(k)}) = Var\, U_l^{(k)}. \tag{6.26}$$

Applying the above to $U_{mn}^{(m)}$ and $U_{\mathcal{D}_{per}}^{(m)}$ under the assumptions of (i) we obtain

$$Var\left(\frac{U_{\mathcal{D}_{per}}^{(m)}}{\left(Var\, U_{\mathcal{D}_{per}}^{(m)}\right)^{1/2}} - \frac{U_{mn}^{(m)}}{\left(Var\, U_{mn}^{(m)}\right)^{1/2}}\right) = 2 - 2\left(\frac{Var\, U_{mn}^{(m)}}{Var\, U_{\mathcal{D}_{per}}^{(m)}}\right)^{1/2} \to 0$$

in view of Theorem 6.3.1, and the result holds via Tchebychev's inequality.

The second part of the theorem follows similarly, since under the assumptions of Theorem 6.3.1 and (ii) we have that $0 < \liminf Var\, U_{mn}^{(m)}$ and in view of (6.21) also $\limsup Var\, U_{\mathcal{D}_{per}}^{(m)} < \infty$. $\qquad\square$

In the context of discussions presented in Chapter 5 we may now appreciate that the proof of the above corollary employs a simplified version of the argument already used in the proof of Theorem 5.2.2.

We conclude this section with the following simple example

Example 6.3.1 (Relative efficiency of the incomplete sample variance). Consider $m = 2$ and a matrix $\{X_{i,j}\}$ of iid real valued random variables with variance $0 < \sigma^2$ and central fourth moment $\mu_4 < \infty$. If we take $h(x, y) = (x - y)^2/2$ then obviously $U_{2n}^{(2)} = S_{(2n)}^2$ is the usual sample variance estimator

$$S_{(2n)}^2 = \sum_{1 \leq i \leq 2, 1 \leq j \leq n} \frac{(X_{ij} - \bar{X})^2}{2n - 1}$$

$$= \frac{1}{2n(2n - 1)} \sum_{\substack{\{(i,j) \neq (k,l) : \\ 1 \leq i, k \leq 2; 1 \leq j, l \leq n\}}} \frac{1}{2}(X_{ij} - X_{kl})^2 \qquad (6.27)$$

where

$$\bar{X} = \frac{1}{2n} \sum_{1 \leq i \leq 2, 1 \leq j \leq n} X_{ij}.$$

Since $r = 1$ in this case, Theorem 6.3.1 applies for the USPD

$$S_{\mathcal{D}_{per}}^2 = \frac{1}{n(n - 1)} \sum_{1 \leq j \neq l \leq n} \frac{1}{2}(X_{1j} - X_{2l})^2$$

which for any given n contains less than a half of the terms present in (6.27).

Let us note that in this case a permanent design is also a minimum variance design since as noted in Example 6.2.1 for a U-statistic of degree $m = 2$ (with square integrable kernel) any balanced design $\mathcal{D} \subseteq \mathcal{S}_{2n,2}$ satisfying $|\mathcal{D}| = 2\binom{2n}{2}$ is meager. A direct comparison of the variances can be also performed, since

$$Var\, S_{(2n)}^2 = \frac{\mu_4 - \sigma^4}{2n} + \frac{\sigma^4}{\binom{2n}{2}}$$

by applying the formula (2.14) in our setting, and

$$Var\, S_{\mathcal{D}_{per}}^2 = \frac{\mu_4 - \sigma^4}{2n} + \frac{\sigma^4}{2\binom{n}{2}}$$

by (6.19) with $m = 2$. As we can see the difference in the above expressions is only in the term of order $O(n^{-2})$. For the sake of example, taking $\sigma^2 = 1$ and $\mu_4 = 3$, we have tabulated below the ratio of $Var\, S_{(2n)}^2/Var\, S_{\mathcal{D}_{per}}^2$ for several different values of n. As we can see from Table 6.1, the efficiency of the permanent design appears reasonable, even for a relatively small sample size.

Table 6.1. Ratios of variances of a complete statistic $S_{(2n)}^2$ and a corresponding USPD for several different values of n.

n	$Var\, S_{(2n)}^2/Var\, S_{\mathcal{D}_{per}}^2$	$\binom{2n}{2}$	% used by PD
5	.89	45	44%
10	.95	190	47%
20	.97	780	49%
50	.99	4950	49.5%

Table 6.2. Relative efficiency of USPD vs complete statistic when m increases.

$m(n)$	approx. sample size $(mn \approx)$	$Var\, T_{mn}^{(m)}/Var\, T_{\mathcal{D}_{per}}^{(m)}$	% used by PD
	10	.938	44.4%
$\ln(n)$	100	.989	21%
	1000	.998	3.69%
	10	.938	44.4%
$\sqrt{n}$	100	.976	2.5%
	1000	.995	0.2%
	10	.938	44.4%
$n/2$	100	.950	0.13%
	1000	.981	$< 10^{-10}$ %

Example 6.3.2 (Relative efficiency of USPD's for the kernels of increasing order).

Let $\mathbb{X}^{(m,n)} = [X_{i,j}]$ $(1 \le i \le m, 1 \le j \le n)$ be a matrix of iid Bernoulli random variables with mean $1/2$. Let us consider the U-statistic $T_{mn}^{(m)}$ based on the kernel $h(x_1, \ldots, x_m) = x_1 \cdots x_m$ and suppose that $m \to \infty$ as $n \to \infty$. Note that

$$T_{mn}^{(m)} = \binom{\sum_{ij} X_{i,j}}{m} / \binom{mn}{m}$$

and the summation above is taken over all the indices i, j. In this case

$$Var\, T_{mn}^{(m)} = \frac{1}{2^m} \sum_{\nu=1}^{m} \frac{\binom{m}{\nu}^2}{\binom{mn}{\nu}}$$

and

$$Var\, T_{\mathcal{D}_{per}}^{(m)} = \frac{1}{2^m} \sum_{\nu=1}^{m} \frac{\binom{m}{\nu}}{\binom{n}{\nu}\nu!}$$

Below we have tabulated several values of the ratio $Var\, T_{mn}^{(m)}/Var\, T_{\mathcal{D}_{per}}^{(m)}$ for three different sequences $m = m(n)$ and different total sample sizes mn. All non-integer values have been rounded to the nearest integer. From Table 6.2 we can clearly see that whereas the efficiency of the incomplete statistic is higher for slowly growing m, the biggest gains in design reduction are achieved for m growing at a faster rate. This is consistent with the observation made earlier.

6.4 Minimal Rectangular Schemes

The overall usefulness of the permanent designs is limited by the fact that they are still fairly large and thus still difficult to compute and, except for some special cases, are less efficient than the random designs of the same size. To address these deficiencies a wide class of "permanent-like" designs will be introduced in this section. These designs, although still based on the rectangular association scheme, are much smaller in size than the "permanent" ones (cf. Definition 6.2.2) but nevertheless share many of the optimal properties of the latter. In particular, some of them are known to minimize the variance over a class of equireplicate designs of the same size and to be uniformly better than the corresponding random designs. However, the price to pay for the reduction in size is that the optimality result can be established across a considerably smaller class of kernels.

We start with the following.

Definition 6.4.1 (Rectangular scheme design). *An equireplicate design* $\mathcal{D} \subseteq \mathcal{S}_{mn,m}$ *is a rectangular scheme (RS) if*

(i) it consists only of the m-sets which are m-transversals of $m \times n$ *matrix as given by (6.17) and*
(ii) for any $i \neq k$ *and* $j \neq l$ *there exists an m-transversal* $\{(1, i_1), \dots, (m, i_m)\}$ $\in \mathcal{D}$ *such that* $\{(i, j), (k, l)\} \subset \{(1, i_1), \dots, (m, i_m)\}$.

Let us note that if we identify the $X_{i,j}$'s with "treatments" and m-subsets of the design with the "blocks" of the classical statistical design theory, then the conditions $(i) - (ii)$ assure that RS is a connected sub-design of some partially balanced, incomplete block (PBIB) design with the rectangular association scheme in the problem of allocating mn treatments into the number of $|\mathcal{D}|$ blocks of size m (see e.g., Street and Street 1987).

The relation between the RS designs and permanent designs is quite obvious. We note first that for $m < n$ the permanent scheme defines partially balanced incomplete block (PBIB) design with three associate classes and in fact is one of the simplest associations schemes leading to such designs (see e.g., Sinha et al. 1993). In a special case of $m = n$ the permanent is equivalent to the celebrated $L_2(n)$ scheme which is PBIB design with two associate classes. Hence, the permanent scheme design, say $\mathcal{D}_{per}$, is always PBIB and also it is a rectangular scheme design which is maximal in the sense that for any rectangular scheme $\mathcal{D} \subset \mathcal{S}_{mn,m}$ we have

$$|\mathcal{D}| \leq |\mathcal{D}_{per}| = \binom{n}{m} m! \qquad (6.28)$$

The RS design-size considerations above motivate the introduction of a notion of a "minimal size" rectangular scheme design. Such a design would consist of a smallest number d of m-transversals needed for the condition (ii) of Definition 6.4.1 to hold. To see what could be a good choice for d let us consider a full set of m-transversals, i.e., the permanent design $\mathcal{D}_{per}$. For this design we have that the number of m-transversals which share a pair of treatments of the form $\{(i, j), (k, l)\}$ $i \neq k$ and $j \neq l$ is simply equal to the number of $(m-2)$-transversals in a $(m-2) \times (n-2)$ matrix of treatments. If we introduce a notation $\mathbb{J}(u, w)$ standing for a rectangular $u \times w$ matrix with all entries equal to one, then the number of such m-transversals is given by

$$Per\,\mathbb{J}(m-2, n-2) = \binom{n-2}{m-2}(m-2)!$$

Hence if we divide the total number of elements in $\mathcal{D}_{per}$ by $Per\,\mathbb{J}(m-2, n-2)$ we obtain the minimal number of m-transversals needed to satisfy the connectedness condition (ii) of Definition 6.4.1 as

$$d = Per\,\mathbb{J}(m, n)/Per\,\mathbb{J}(m-2, n-2) = n(n-1). \tag{6.29}$$

Note that this derivation leads to designs in which each pair of transversals share at most one element. This motivates the following definition.

Definition 6.4.2 (Minimal rectangular scheme). *Any RS design $\mathcal{D}$ such that for each pair of its m-transversals there exists at most one common element is called* minimal.

Note that by the argument preceding the definition it follows that if $\mathcal{D}$ is a minimal rectangular scheme then $|\mathcal{D}| = n(n-1)$. Alternatively, it follows from (6.16). Unfortunately, for an arbitrary integer n such minimal designs may not exist. We defer the discussion of that issue to the next section. However, it turns out that when they do exist, MRS designs have some optimal properties.

Theorem 6.4.1. *MRS design is a minimum variance design (MVD), that is, has the smallest variance in the class of all the equireplicate designs (not necessarily rectangular) $\mathcal{D} \subseteq \mathcal{S}_{mn,m}$ satisfying $|\mathcal{D}| = n(n-1)$. Moreover, its variance is*

$$V_0 = \frac{m}{n}Var\,g_1 + \frac{1}{n(n-1)}\sum_{c=2}^{m}\binom{m}{c}Var\,g_c$$

or, via the relation (6.3),

$$V_0 = \frac{(n-2)m}{n(n-1)}Var\,h_1 + \frac{1}{n(n-1)}Var\,h_m$$

Proof. By the variance formula (6.7) it follows that for fixed m, n it is enough to minimize the quantities

$$B_\nu = \sum_{c=\nu}^{m} f_c\binom{c}{\nu} = \sum_{S_\nu \in \mathcal{S}_{mn,\nu}} n^2(S_\nu)$$

for $\nu = 2, \ldots, m$, since for all equireplicate designs $n(S_1) = s$ is constant for any $S_1 \in \mathcal{S}_{mn,1}$, thus by (6.16) we have $B_1 = mn\,s^2 = md^2/n$.

Let us note that for all designs of the same size $d = n(n-1)$ we have

$$B_\nu \geq \sum_{S_\nu \in S_{mn,\nu}} n(S_\nu) = n(n-1)\binom{m}{\nu} = A_\nu$$

where the last equality is a definition. But for MRS design we have $n(S_2) \leq 1$ which implies $n(S_\nu) \leq 1$ and hence $B_\nu = A_\nu$ for $\nu = 2,\ldots,m$. Once we have identified the B_ν's the final variance formula for MRS design follows immediately from (6.7).   □

We shall refer to the incomplete U statistics based on MRS designs as U-statistics of minimal rectangular scheme or UMRS's. Let us note that the UMRS's (unlike USPD) are uniformly more efficient than the corresponding ones based on random designs, that is, the designs in which the $n(n-1)$ of the m-sets belonging to the design are selected at random from $S_{mn,m}$ (with or without replacement). To that end we shall need first the following result.

Lemma 6.4.1 (Variance formula for U-statistics of random design).
Consider an incomplete U-statistic (6.5) where the design $\mathcal{D}$ is created by selecting at random $|\mathcal{D}| = d$ subsets from $S_{l,k}$. Then the variance $V(RND)$ of such a U-statistic is as follows.

(i) For selection with replacement

$$V(RND) = \left(1 - \frac{1}{d}\right) Var\, U_l^{(k)} + \frac{1}{d} Var\, h_k. \tag{6.30}$$

(ii) For selection without replacement

$$V(RND) = \frac{K(d-1)}{d(K-1)} Var\, U_l^{(k)} + \frac{K-d}{d(K-1)} Var\, h_k$$

where $K = \binom{l}{k}$.

Proof. We proof only (i) as (ii) is similar. If $S_1,\ldots,S_d$ are d subsets selected from K subsets at random with replacement then for $i \neq j$,

$$Cov\,(h(S_i), h(S_j)) = K^{-2} \sum_{S \in S_{l,k}} \sum_{T \in S_{l,k}} Cov\,(h(S), h(T))$$

$$= Var\, U_l^{(k)}.$$

Thus

$$V(RND) = d^{-2}\left(\sum_{i \neq j} Cov\,(h(S_i), h(S_j)) + \sum_i Var\, h(S_i)\right)$$

$$= d^{-2}\left(d(d-1)Var\, U_l^{(k)} + d\, Var\, h_k\right).$$

□

Using the above variance formula for the random design U-statistic for selection with replacement in case when we select $d = n(n-1)$ m-sets from $S_{mn,m}$ we obtain from (6.30)

$$V(RND) = \left(1 - \frac{1}{n(n-1)}\right) Var\, U_{mn}^{(m)} + \frac{1}{n(n-1)} Var\, h_m.$$

If we now denote by V_0 the variance of UMRS then for $n \geq 2$ we have, by (6.3) and Theorem 6.4.1 above that

$$V(RND) - V_0 =$$

$$\left(1 - \frac{1}{n(n-1)}\right) \sum_{c=2}^{m} \frac{\binom{m}{c}^2}{\binom{mn}{c}} Var\, g_c + \frac{m}{n^2} Var\, g_1 > 0.$$

A similar relation holds for selection without replacement if only $m, n \geq 3$.
    Recall from the previous section that

$$ARE = \lim_{l \to \infty} \frac{Var\, U_l^{(k)}}{Var\, U_D^{(k)}}$$

and that in Theorem 6.3.1 it was shown that for the permanent design U-statistics in a large class of kernels their ARE was equal unity. Interestingly enough, similar result, although for a smaller class and only for non-degenerate kernels, is also true for UMRS's. Hence under certain conditions the variance of an incomplete U-statistic based on MRS for which $d = n(n-1)$ may be close to that of a corresponding U-statistic of permanent design for which $d = \binom{n}{m} m!$ or a complete (non-degenerate) U-statistic (6.15) for which $d = \binom{mn}{m}$. This turns out to be the case in a class of kernels, where the corresponding variances of the conditional kernels grow not too fast relatively to each other. In order to state and prove this result we shall need the following sharp lower bound on the variance of a U-statistic, due to Vitale (1992).

Lemma 6.4.2. *Suppose that $U_l^{(k)}$ is a U-statistic of level of degeneration $r-1$ based on the kernel of k arguments h_k satisfying $E\, h_k^2 < \infty$, then*

$$\frac{\binom{k}{r}^2}{\binom{l}{r}} Var\, h_r \leq Var\, U_l^{(k)}.$$

$\square$

We do not give the proof of this result here as its relies on some linear programing methods which are outside our scope. Interested reader is referred to the original paper Vitale (1992, theorem 6.2). It turns out that the lemma above is useful in proving the following.

Theorem 6.4.2. *Let us consider any non-degenerate U-statistic $U_{mn}^{(m)}$ given by (6.15) and let the variance of UMRS be denoted by V_0 and $\Delta_k^{(n)}$ stand for the ratio $Var\, h_k/(k\, Var\, h_1)$ for $k = 1, \ldots, m$. Then we have*

$$\frac{V_0}{Var\, U_{mn}^{(m)}} - 1 \le \frac{(n+1)\,(\Delta_m^{(n)} - 1) - (m-1)^2\,(\Delta_2^{(n)} - 1)}{n^2 - 1 + (m-1)^2\,(\Delta_2^{(n)} - 1)} \le \frac{\Delta_m^{(n)} - 1}{n - 1}.$$

In particular, if

$$\limsup_n \Delta_m^{(n)} < \infty \tag{6.31}$$

then ARE of UMRS vis à vis $U_{mn}^{(m)}$ equals one.

Via (1.18) it follows that $\Delta_k^{(n)} \ge 1$ for $k \ge 1$ hence the second inequality above is immediate. The condition (6.31) requires the entries of the double array $\Delta_k^{(n)}$ for $k = 1, \ldots, m$ and $n = 1, 2, \ldots$ to be uniformly bounded from above and may be thought of as the restriction on the rate of growth of the variance of $U_{mn}^{(m)}$, since it implies by (6.4) that $Var\, U_{mn}^{(m)} = O(\frac{m}{n}\, Var\, h_1)$. Let us note that this condition is trivially satisfied for any U-statistic for which the kernel function does not depend upon the sample size.

Proof. Let us assume (without loss of generality) that $E\, h_m = 0$. Denoting as before the variance of UMRS by V_0 we consider the ratio

$$0 \le \frac{V_0}{Var\, U_{mn}^{(m)}} - 1 = \frac{\frac{(n-2)m}{n(n-1)}\, Var\, h_1 + \frac{1}{n(n-1)}\, Var\, h_m}{\frac{m}{n}\, Var\, h_1 + Var\, R_{mn}^{(m)}} - 1$$

where the last expression is obtained by applying the form (1.16) of the H-decomposition to $U_{mn}^{(m)}$ (as $r = 1$) and using the variance formula for V_0 derived in Theorem 6.4.1. Let us note that we can obtain the ratio's upper bound as follows

$$\frac{\frac{(n-2)\,m}{n(n-1)}\, Var\, h_1 + \frac{1}{n(n-1)}\, Var\, h_m}{\frac{m}{n}\, Var\, h_1 + Var\, R_{mn}^{(m)}} - 1 \le \frac{\frac{m\, Var\, h_1}{n(n-1)}\left(\Delta_m^{(n)} - 1\right) - Var\, R_{mn}^{(m)}}{\frac{m}{n}\, Var\, h_1 + Var\, R_{mn,1}}.$$

Recall that $R_{mn}^{(m)}$ is itself a U-statistic with a level of degeneration equal to one since it is based on the kernel $\tilde{h}_m(x_1, \ldots, x_m) = h_m(x_1, \ldots, x_m) - \sum_{i=1}^{m} h_1(x_i)$. By means of Vitale's result (Lemma 6.4.2) we obtain,

$$Var\, R_{mn}^{(m)} \ge \binom{m}{2}^2 \binom{mn}{2}^{-1} Var\, \tilde{h}_2 = \binom{m}{2}^2 \binom{mn}{2}^{-1} (Var\, h_2 - 2Var\, h_1)$$

$$\ge \frac{m(m-1)^2}{n(n^2-1)}\, Var\, h_1 \left(\Delta_2^{(n)} - 1\right).$$

Combining the above inequalities entails the final result. □

We should perhaps note that the right-most inequality in the assertion of the theorem can be inferred directly from Vitale's result with $r = 1$. The intermediate inequality derived above is, however, relevant when estimating relative efficiency with fixed sample sizes. Let us also note that the result indicates that under the proviso (6.31) the ratio of the variances converges to one (from above) at the rate of $O(1/n)$, regardless of the size m of the kernel. This implies, in particular, that UMRS's should be equally efficient in the case when kernel's order m grows along with the sample size.

6.5 Existence and Construction of MRS

Let us note that, as mentioned before, the problem of finding MRS can be formulated in terms of the design theory as the search for a binary, partially balanced, incomplete block design with at most three associate classes (or PBIB(3) design) following the rectangular (association) scheme with the following design parameters: number of treatments $v = mn$, block size $t = m$, number of blocks $b = (n-1)\,n$, and number of replications $s = n-1$. In order to construct such MRS as well as answer the question of its existence let us first discuss some additional concepts needed in the sequal.

6.5.1 Strongly Regular Graphs

We first introduce two useful notions, namely that of a regular and a strongly regular graph.

Definition 6.5.1 (Regular graph). *Let V be a set of vertices and $E = \{(i_k, j_k) : i_k, j_k \in V; i_k \neq j_k\}$ be a set of edges. A pair $G = (V, E)$ is called an undirected graph of order $d = |V|$. It is called a* regular *(undirected) graph of order d and degree k if for each vertex $v \in V$ the number of edges containing v is constant and equals k.*

A simple example of a regular graph is a complete graph $K(d) = G(V, E_{max})$, i.e. a graph for which the set of edges is $E_{max} = \{(i, j) : i, j \in V; i \neq j\}$. Recall that we call two vertices *adjacent* if they share an edge. We shall denote the adjacency matrix of G by $\mathbb{A}$. It turns out that the following special subclass of regular graphs shall be especially useful for our purpose.

Definition 6.5.2 (Strongly regular graph). *A strongly regular graph G with a set of the parameters (d, k, λ, μ) is a regular graph of order d and degree k such that for any $a, b \in V$ $a \neq b$*

(i) *if a, b are adjacent vertices there exist exactly λ further vertices of G adjacent to both a and b;*

(ii) *if a, b are non-adjacent vertices there exist exactly μ further vertices adjacent to both a and b.*

An example of a strongly regular graph is given in Figure 6.1. For a more detailed introduction to the properties and basic theory of strongly regular graphs see, e.g., Brualdi and Ryser (1991).

Definition 6.5.3 (Complement of a graph). *Let E_{max} denote a complete set of edges, i.e, the set created by considering all possible pairs of vertices in G. The graph $\bar{G} = (V, E_{max} \backslash E)$ is called a complement of G.*

The following two properties of strongly regular graphs shall be subsequently of interest.

Lemma 6.5.1. *Let G be a strongly regular graph with parameters (d, k, λ, μ). Then*

(i) $\bar{G}$ is also a strongly regular graph with parameters $(\bar{d} = d, \bar{k} = d - k - 1, \bar{\lambda} = d - 2k - 2 + \mu, \bar{\mu} = d - 2k + \lambda)$.
(ii) If $\mu = 0$ then G is a union of disjoined complete graphs $K(k+1)$.

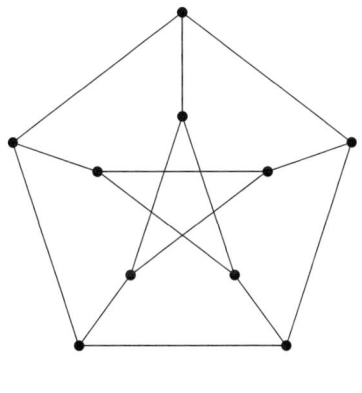

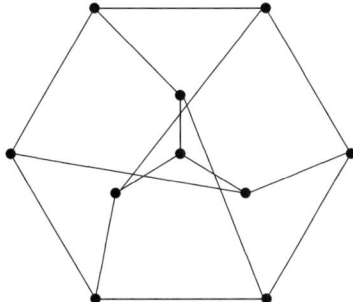

Fig. 6.1. The five- and three-symmetric representations of the Petersen graph which is a classical example of a strongly regular graph with parameters $(10, 3, 0, 1)$.

Proof. We first prove part (i). Let $G(d, k, \lambda, \mu)$ be a strongly regular graph with adjacency matrix $\mathbb{A}$. Then the adjacency matrix determines uniquely $G(d, k, \lambda, \mu)$. We know that the entry in the (i, j) position of $\mathbb{A}^2$ equals the number of paths of length two with vertices i and j as endpoints. This number however, by the definition of a strongly regular graph, has to be equal to k, λ, or μ depending on whether the vertices are equal, adjacent or nonadjacent. If we let as before $\mathbb{J} = \mathbb{J}(d, d)$ denote a $d \times d$ matrix of unit entries, then we see that the above argument implies that G is a strongly regular graph with given parameters if and only if its adjacency matrix $\mathbb{A}$ satisfies

$$\mathbb{A}^2 = k\mathbb{I} + \lambda\mathbb{A} + \mu(\mathbb{J} - \mathbb{I} - \mathbb{A}). \tag{6.32}$$

Note that the adjacency matrix of $\bar{G}$ is simply $\bar{\mathbb{A}} = \mathbb{J} - \mathbb{I} - \mathbb{A}$. Consider now

$$(\mathbb{J} - \mathbb{I} - \mathbb{A})^2 = \mathbb{J}^2 - 2\mathbb{J} - \mathbb{J}\mathbb{A} - \mathbb{A}\mathbb{J} + \mathbb{I} + 2\mathbb{A} + \mathbb{A}^2. \tag{6.33}$$

Due to (6.32) and since $\mathbb{J}^2 = d\mathbb{J}$ and $\mathbb{J}\mathbb{A} = \mathbb{A}\mathbb{J} = k\mathbb{J}$, the expression on the right hand side of (6.33) may be written as

$$(d - 2 - 2k)\mathbb{J} + \mathbb{I} + 2\mathbb{A} + k\mathbb{I} + \lambda\mathbb{A} + \mu(\mathbb{J} - \mathbb{I} - \mathbb{A})$$
$$= (d - k - 1)\mathbb{I} + (n - 2 - 2k + \mu)(\mathbb{J} - \mathbb{I} - \mathbb{A}) + (d - 2k + \lambda)\mathbb{A}$$

which proves (i).

In order to prove (ii) we multiply the equation (6.32) by a column vector of unit values to obtain

$$k^2 = k + \lambda k + \mu(d - 1 - k).$$

For $\mu = 0$ this implies that
$$\lambda = k - 1$$

and thus each vertex of G belongs to a complete graph on $k + 1$ vertices $K(k + 1)$. Hence part (ii) follows.   □

6.5.2 MRS and Orthogonal Latin Squares

As noted earlier even though MRS designs have some optimal properties so far we have not shown that they may be in fact constructed. Before we turn to this problem, let us first introduce the following notion.

Definition 6.5.4 (Mutually Orthogonal Latin Squares). *A Latin square of order n is an $n \times n$ array based on a set T of n symbols, such that each row and each column contains each symbol exactly once. Two Latin squares $L_1 = (a_{ij})$ and $L_2 = (b_{ij})$ of order n on symbol set T are said to be orthogonal if every element in $T \times T$ occurs exactly once among the n^2 pairs (a_{ij}, b_{ij}), $1 \leq i, j \leq n$. A set of Latin squares $L_1, L_2, L_3, \cdots, L_k$ is mutually orthogonal, or a set of MOLS's, if for every $1 \leq i < j \leq k$, L_i and L_j are orthogonal.*

Let $N(n)$ denote the maximal number of Latin squares in a set of MOLS's of side n. It is easily verifiable that the total number of MOLS's of order n cannot exceed $n-1$, i.e., $N(n) \leq n-1$. If $N(n) = n-1$, then a set of MOLS's is said to be *complete*. For more details on the properties of Latin squares we refer to Street and Street (1987, p 126–133).

We are now ready to prove the main result of this section which relates the existence of MRS with arbitrary $m \leq n$ to the existence of a complete set of MOLS's. The method of the proof gives also an explicit construction of the design from a complete set of MOLS's.

Theorem 6.5.1. *A minimal rectangular scheme design having parameters $v = mn$, $b = n(n-1)$, $s = n-1$, $t = m$, exists for each $m = 2, 3 \ldots, n$ if and only if there exists a set of $n-1$ mutually orthogonal $n \times n$ Latin squares.*

Proof. We first show sufficiency. For any $2 \leq m \leq n$ consider mn treatment (sample) indices arranged in a rectangular array of m rows and n columns and consider a complete set of $n-1$ MOLS's of order n. Superimposing any fixed m rows of any two these Latin squares on the rectangular array of mn indices and forming blocks with treatment indices corresponding to the same symbol on the superimposed Latin square, we can generate $n(n-1)$ blocks. Clearly $v = mn$, $b = n(n-1)$, $s = n-1$, $t = m$ and the completeness of a set of MOLS's implies other MRS conditions as well.

For the proof of necessity we take $m = n$ and consider a collection of $b = n(n-1)$ blocks of $t = n$ treatments belonging to MRS design. We note that by definition of MRS any two distinct blocks have at most one treatment in common. Moreover, since every treatment occurs exactly $s = n-1$ times in the design it follows that for every block there is exactly $n(n-2)$ different blocks which share one of its n treatments. Let us introduce a square $0-1$ matrix $C = [c_{ij}]$ of $n(n-1)$ rows and columns with the off-diagonal entries c_{ij} being one or zero depending on whether the blocks i and j share a treatment. We put $c_{ii} = 0$. The symmetric matrix C is know in design theory as *the concurrence matrix of a dual design* and has the property that its rows and columns sum up to $n(n-2)$. It is easy to see that the matrix C defines a strongly regular graph of degree $k = n(n-2)$ on $d = n(n-1)$ vertices with parameters $\lambda = n(n-3)$ and $\mu = n(n-2)$. Moreover, a complement of this graph has parameters $\bar{d} = d, \bar{k} = n-1, \bar{\lambda} = n-2, \bar{\mu} = 0$ and thus by Lemma 6.5.1 is a disjoint union of complete graphs $K(n)$ which implies that the MRS design can be partitioned into $n-1$ disjoint sets of n non-overlapping blocks of length n. Let us note that every such set must necessarily contain all of the $v = n^2$ treatments. Having found the partitioning of the design blocks into $n-1$ sets of n blocks, we see that the properties of MRS design now impose that this partitioning must correspond to a collection of $n-1$ mutually orthogonal $n \times n$ Latin squares. Indeed, consider an arbitrary set in the partitioning. By the properties of the MRS design it has to consist of the n blocks which are disjoint n-transversals of the $n \times n$ matrix of treatments.

Hence associating each n-transversal with a different symbol from the set $\{1, \ldots, n\}$ we note that no two such symbols may share a row or a column in the matrix of treatments, which is a Latin square condition. Thus we see that each set of the partitioning corresponds to a Latin square. The fact that these $n - 1$ Latin squares of size n must be orthogonal follows immediately from the fact that no treatments may be shared by any blocks of the MRS design more than once. Thus the partitioning gives us a complete set of MOLS's of order n on the set of symbols $\{1, 2, \ldots, n\}$.   □

In general, the existence of a complete set of MOLS for an arbitrary integer $n \geq 3$ is an intriguing combinatorial problem dating back to the celebrated Euler's conjecture (see, e.g., Bose et al. 1960). However, for the set of integers of the form $n = p^s$ where p is prime and $s \geq 1$, it is known that $N(n) = n - 1$. Although we shall not discuss it here, it turns out that the construction of a corresponding set of MOLS's is quite simple (see, for instance, Street & Street 1987, pp. 129–131). In view of this we have the following more explicit result on the existence of MRS designs.

Corollary 6.5.1. *If $n = p^i$ for some positive integers i, p, where $p > 1$ is also a prime number, then MRS design with parameters $v = mn$, $b = n(n-1)$, $s = n - 1$, $t = m$, exists for each $m = 2, 3 \ldots, n$.*   □

A rectangle created by removing last $n - m$ row(s) from a Latin square is called $m \times n$ *Latin rectangle*. Theory of mutually orthogonal Latin rectangles (cf., e.g., Horák et al. 1997) parallels that of Latin squares. Using the notion of Latin rectangles one may state a slightly more general version of the result in Theorem 6.5.1.

Theorem 6.5.2. *A minimal rectangular scheme design having parameters $v = mn$, $b = n(n - 1)$, $s = n - 1$, $t = m$, exists for a fixed m satisfying $2 \leq m \leq n$ if and only if there exists a set of $n - 1$ orthogonal $m \times n$ Latin rectangles.*

Proof. The proof of this result is very similar to that of Theorem 6.5.1 and hence omitted.   □

6.6 Examples

Let us conclude this chapter with some examples of the applicability of the results discussed in the previous section.

Example 6.6.1. Let $n = 6$ then it is well known (see. e.g., Stinson 1984) that $N(6) = 1$. Hence by Theorem 6.5.1 no MRS design exists for the parameters $v = 36, b = 30, t = 6, s = 5$.

On the other hand, for n being a prime power we can always construct a set of $n - 1$ MOLS's and thus also the MRS design.

Example 6.6.2. Consider $T = \{a, b, c, d\}$ and $n = 4 = 2^2$, so there are $n - 1 = 3$ MOLS's of order $n = 4$ as shown below (for the algorithm to construct these MOLS's see, e.g., Street & Street 1987):

$$
\begin{array}{cccc\ cccc\ cccc}
a & b & c & d & a & b & c & d & a & b & c & d \\
d & c & b & a & c & d & a & b & b & a & d & c \\
b & a & d & c & d & c & b & a & c & d & a & b \\
c & d & a & b & b & a & d & c & d & c & b & a
\end{array}
$$

Consider $m = 4$. Superimposing the following 4×4 array of sample indices on the rows of each of the above Latin squares

$$
\begin{array}{cccc}
1 & 2 & 3 & 4 \\
5 & 6 & 7 & 8 \\
9 & 10 & 11 & 12 \\
13 & 14 & 15 & 16
\end{array}
$$

we see that, for instance, the symbol a corresponds to the index '1' in the first row, the index "8" in the second row, the index '10' in the third row, and the index '15' in the fourth row which thus gives us the block $\{1, 8, 10, 15\}$. Following the same procedure for the symbols b, c, d, we obtain the following required rectangular design with parameters $v = mn = 16$, $b = n(n - 1) = 12$, $t = m = 4$, $s = 3$ and the design blocks displayed by columns below:

$$
\begin{array}{cccc\ cccc\ cccc}
1 & 2 & 3 & 4 & 1 & 2 & 3 & 4 & 1 & 2 & 3 & 4 \\
8 & 7 & 6 & 5 & 7 & 8 & 5 & 6 & 6 & 5 & 8 & 7 \\
10 & 9 & 12 & 11 & 12 & 11 & 10 & 9 & 11 & 12 & 9 & 10 \\
15 & 16 & 13 & 14 & 14 & 13 & 16 & 15 & 16 & 15 & 14 & 13
\end{array}
$$

Having constructed the design for $m = 4$ we can now easily construct designs for any $m \leq 4$ by simply deleting the appropriate rows of the above design matrix. Thus, for instance, the MRS design for $m = 3$ may be obtained be deleting the last row, for $m = 2$ by deleting the two last rows, etc.

In the final example, let us make a comparison of the relative efficiency between MRS's and random designs for a specific U-statistic.

Example 6.6.3. Let us consider a particular U-statistic of the form (6.15)

$$
H_{mn}^{(m)} = \binom{mn}{m}^{-1} \sum_{(i_1,\ldots,i_m):\{i_1,\ldots,i_m\}\subset\{1,\ldots,n\}} \max(X_{1,i_1}, \ldots, X_{m,i_m}).
$$

Let F be the underlying distribution function for the $X_{i,j}$'s and let X be equidistributed with $X_{i,j}$'s. It is not difficult to see that the above statistic is an unbiased estimator of the so called *probability weighted moment* $\beta_m = m\, E\,[X\, F(X)^{m-1}]$, that is, $E\, H_{mn}^{(m)} = \beta_m$. The sequence of constants $(\beta_m)_{m \geq 1}$ characterizes the underlying distribution, in the sense that for different F's we shall have different sequences $(\beta_m)_{m \geq 1}$. For the sake of computational simplicity let us assume henceforth that F is a distribution function of the uniform distribution $U(0,1)$. Then, we have for $c = 1, \ldots, m$

$$h_c(x_1, \ldots, x_c) = E \max(x_1, \ldots, x_c, X_{c+1}, \ldots, X_m)$$
$$= \frac{m - c}{m - c + 1} + \frac{\max(x_1, \ldots, x_c)^{m-c+1}}{m - c + 1},$$

where the X_k's are mutually independent and equidistributed with X, as well as

$$Var\, h_c = \frac{c}{(m + 1)^2\,(2m - c + 2)}.$$

From the above it follows that T_m is non-degenerate ($r = 1$) and satisfies the condition (6.21), since for $m \geq 1$

$$\frac{Var\, h_m}{m\, Var\, h_1} = \frac{2m + 1}{m + 2} \leq 2.$$

For the statistic $H_{mn}^{(m)}$, let us make a comparison of the efficiency between MRS design and the one based on the random selection method (when selection is done without replacement). If we denote

$$RE_{MRS} = \frac{Var\,(H_{mn}^{(m)})}{V_0} \quad \text{and} \quad RE_{RND} = \frac{Var\,(H_{mn}^{(m)})}{V(RND)}$$

where $V(RND)$ is given by (6.4) and V_0 is the variance for MRS design as given in Theorem 6.4.1 we find out that, for instance, for the parameters considered in the previous example we have the values summarized in Table 6.3 below.

As we can see from Table 6.3, MRS seems to have a distinct edge over the random design selection for small sample sizes, which is consistent with the above derivations (very similar results are also obtained if we consider selection without replacement).

In order to compare the quantities RE_{MRS} and RE_{RND} for large kernel and sample sizes we have tabulated several values of the variance ratios for the three different sequences $m_n = m(n)$ and different total sample sizes mn. All non-integer values have been rounded to the nearest integer. The results are presented in Table 6.4. From that table we can clearly see that MRS seems to

Table 6.3. Relative efficiency of MRS vs random design RND for fixed m.

m	sample size $(mn =)$	RE_{MRS}	RE_{RND}
2	8	0.956	0.758
3	12	0.937	0.737
4	16	0.924	0.724

Table 6.4. Relative efficiency of MRS vs random design RND when m increases.

$m(n)$	approx. sample size $(mn \approx)$	RE_{MRS}	RE_{RND}
$\ln(n)$	10	0.898	0.816
	100	0.993	0.953
	1000	0.999	0.992
$\sqrt{n}$	10	0.960	0.776
	100	0.988	0.933
	1000	0.997	0.982
$n/2$	10	0.957	0.773
	100	0.977	0.894
	1000	0.990	0.960

be more efficient than random selection also for larger sample sizes, especially when m is growing at a faster rate. For very large sample sizes (1000) the differences are less pronounced but also clearly visible. The appeal of the MRS design seems to be, therefore, in its efficiency and also in the relative simplicity of its construction which in essence requires only the creation of the appropriate set of Latin squares, a task which can be easily done, for instance, by a computer program.

6.7 Bibliographic Details

The original article by Blom (1976) as well as Lee (1982) and Lee (1990, chapter 4) are good sources of information on incomplete U-statistics for the case of both random and deterministic selection of subsets. In particular, the introductory material herein on incomplete U-statistics is largely taken from chapter 4 in Lee (1990). We refer readers there for further references and examples of various applications (including e.g., applications to the analysis of random graphs). Additionally, the papers of Frees (1989), Politis and Romano (1994) and Nowicki and Wierman (1988)) as well as Janson (1984) provide some useful results on the asymptotic behavior of incomplete U-statistics with random subsets selection. Some interesting non-random designs are also presented in Brown and Kildea (1978), Enqvist (1985), Herrndorf (1986). The material on permanent designs presented in this chapter is taken from Rempała and Wesołowski (2003) whereas the discussion of minimal rectangular designs and the result of Theorem 6.5.1 come from the original paper of Rempała and Srivastav (2004). The section on strongly regular graphs is largely based on chapter 5 in Brualdi and Ryser (1991). The general reference to experimental design theory is provided e.g., in the monograph by Street and Street (1987). The methods of constructing designs based on rectangular schemes of associations have been considered by many authors, typically, however, not in the context of U-statistics and for different sets of parameters than MRS. For some related PBIB(3) designs with rectangular association schemes one can see, for instance, the review paper by Sinha and Mitra (1999) and references therein, as well as, Sinha et al. (1993), Bagchi (1994), Vartak and Diwanji (1989) and earlier Stahly (1976).

In the special case of $m = n$ the MRS design is a PBIB(2) design of $L_2(n)$ association type and its construction is discussed e.g., in Raghavarao 1988. It follows from the general result on PBIB(2) designs of Cheng and Bailey (1991) that for $m = n$ the MRS and its dual are both optimal in the appropriate class of equireplicate designs with respect to a large class of combinatorial criteria including the so-called A, D and E criteria. In particular, Bagchi and Cheng (1993) showed the E-optimality of the MRS design for $m = 2$ (note that in this case it is also a permanent design).

7

Products of Partial Sums and Wishart Determinants

7.1 Introduction

In this chapter we shall consider a problem of asymptotic behavior of products of partial sums of identically distributed random variables and also products of U-statistics.

The results discussed here shall eventually lead us to a limit theorem for determinants of Wishart matrices as well as several extensions of the classical result on the lognormal limit for the products of independent, identically distributed, positive, and square integrable random variables which derive from the classical central limit theorem (CLT) as stated for instance in Corollary 4.3.1. In the present chapter, we show that the limiting laws of the products of subsequent partial sums of independent and equidistributed random variables are also lognormal and that this result extends to non-degenerate "classical" U-statistics with fixed kernel functions.

Rather unexpectedly, one of the first results for the products of partial sums of independent identically distributed random variable appeared during investigations of the limiting properties of sums of records. We recall that for a sequence (X_n) of random variables the sequence of records is defined in the following way. Let $T_1 = 1$ be the first record time and $R_1 = X_1$ be the first record. Then for $n > 1$ the n-th record time T_n and the n-th record R_n are defined recursively as

$$T_n = \inf\{j > T_{n-1} : X_j > R_{n-1}\} \quad \text{and} \quad R_n = X_{T_n}.$$

If X_n's are independent and have the common distribution function F, which is continuous, then the following in-distribution representation occurs

$$(R_n)_{n \geq 1} \overset{d}{=} (\psi_F(S_n))_{n \geq 1},$$

where $S_n = Z_1 + \ldots + Z_n$ is the n-th partial sum of the sequence (Z_n) of independent standard (i.e. with the unit mean) exponential random variables, and $\psi_F(x) = F^{-1}(1 - e^{-x})$, $x \geq 0$. Here F^{-1} denotes the quantile function (the generalized inverse of the distribution function) defined as $F^{-1}(w) = \sup\{x : F(x) \leq w\}$, $w \in [0, 1]$. Thus for the n-th sum of records the following in-distribution representation holds

$$\sum_{k=1}^{n} R_k \stackrel{d}{=} \sum_{k=1}^{n} \psi_F(S_k).$$

If independent random variables in the sequence (X_n) have the common Gumbel distribution, i.e.

$$F(x) = 1 - e^{-e^{x}}, \quad x \in \mathbf{R},$$

then $\psi_F(x) = \log(x)$, $x > 0$, and thus

$$\sum_{k=1}^{n} R_k \stackrel{d}{=} \log\left(\prod_{k=1}^{n} S_k\right).$$

Investigations of the asymptotic behaviour of sums of records of Gumbel sequences led to the following version of the CLT for a sequence (Z_n) of independent and identically distributed exponential random variables with the mean equal one

$$\frac{\sum_{k=1}^{n} \log(S_k) - n \log(n) + n}{\sqrt{2n}} \stackrel{d}{\to} \mathcal{N} \tag{7.1}$$

as $n \to \infty$, where $S_k = Z_1 + \ldots + Z_k$, $k = 1, 2, \ldots$, and $\mathcal{N}$ is a standard normal random variable. The original proof of this result was heavily based on a very special property of exponential (gamma) distributions: namely on joint independence of ratios of subsequent partial sums and the last sum. Also the classical result on weak limits for records was exploited.

Observe that, via the Stirling formula, the relation (7.1) can be equivalently stated as

$$\left(\prod_{k=1}^{n} \frac{S_k}{k}\right)^{\frac{1}{\sqrt{n}}} \stackrel{d}{\to} e^{\sqrt{2}\mathcal{N}}.$$

The main purpose of our current discussion is to give a wider exposition of the topic of limits of products of random sums and in particular to extend the limiting result recalled above to any sequence of sums of square integrable, positive, independent and identically distributed random variables.

Let $\mathbb{X}^{(\infty)}$ be a random matrix of square integrable entries satisfying as before the assumptions (A1-A2). We modify it by replacing all the above diagonal entries by zeros. In our current scenario instead of the P-statistic we consider products of the row-sums of matrix entries.

Herein we shall only consider two simplest special cases of the general exchangeability assumptions (A1-A2). The first one is $\mathbb{X}^{(\infty)} = \mathbb{X}$, i.e., when

$\underline{X}$ is one dimensional projection matrix and the second one is when $\mathbb{X}^{(\infty)}$ is a matrix of all independent identically distributed entries. In the both cases we modify $\mathbb{X}^{(\infty)}$ by inserting zeroes above the diagonal.

Note that the relation (7.1) for the exponential variables above was obtained under the assumption $\mathbb{X}^{(\infty)} = \underline{X}$. This assumption obviously implies that the sequence of partial sums $(S_k)_{k \geq 1}$ is obtained from a single sequence of independent random variables (and thus the entries of the sequence $(S_k)_{k \geq 1}$ are not independent). The extensions of (7.1) to arbitrary square integrable random variables as well as to U-statistics, is discussed in the next two sections.

In case when $\mathbb{X}^{(\infty)}$ consists of all independent entries, we obtain a triangular array of independent and identically distributed random variables and consider products of its subsequent row sums. This second setting leads to independent partial sums and, as it is pointed out in the last section of the chapter, allows in particular to investigate the limiting behavior of determinants of Wishart matrices. The result on random determinant for Wishart matrices (given below in Theorem 7.4.1) is on one hand an interesting application of the methods developed and discussed in this chapter, but on the other hand it also allows for a comparison of a limiting behavior of a random determinant with that of a random permanent as presented in Chapter 3.

Note that an alternative setting is by considering a single sequence (X_n) of independent identically distributed random variables and a sequence of sums

$$S_k = X_{a_k+1} + X_{a_k+2} + \ldots + X_{a_{k+1}},$$

where $a_k = k - 1$, $k = 1, 2, \ldots$, in the first setting of subsequent partial sums, and $a_k = \frac{k(k-1)}{2}$, $k = 1, 2, \ldots$, in the second setting of independent sums of growing length.

The limiting behaviour of products of sums arising form arbitrary sequences (a_k) is not known in general. The problem can be further generalized by introducing the parameters (r_k) for lengths of sums. In the present setting $r_k = k$.

7.2 Products of Partial Sums for Sequences

As outlined above, let us first consider a sequence $(X_n)_{n \geq 1}$ of independent and identically distributed random variables, along with a corresponding sequence of their partial sums $(S_k)_{k \geq 1}$ The main result we would like to obtain is a general version of (7.1), without assumptions of any particular distribution for the X_i's.

Theorem 7.2.1. *Let (X_n) be a sequence of independent and identically distributed positive square integrable random variables. Denote $\mu = EX_1 > 0$,* *the coefficient of variation* $\gamma = \sigma/\mu$, *where* $\sigma^2 = Var(X_1)$, *and* $S_k = X_1 + \ldots + X_k$, $k = 1, 2, \ldots$. *Then*

$$\left(\frac{\prod_{k=1}^{n} S_k}{n! \mu^n} \right)^{\frac{1}{\gamma \sqrt{n}}} \xrightarrow{d} e^{\sqrt{2}\mathcal{N}} , \qquad (7.2)$$

where $\mathcal{N}$ *is a standard normal random variable.*

Before proving the above result we will establish a version of the classical central limit theorem, essentially, for scaled independent and identically distributed random variables. To this end we will use the CLT for triangular arrays (see, Corollary 4.3.1), so the basic step in the proof will rely on verifying the Lindeberg condition.

Lemma 7.2.1. *Under the assumptions of Theorem 7.2.1*

$$\frac{1}{\gamma \sqrt{2n}} \sum_{k=1}^{n} \left(\frac{S_k}{\mu k} - 1 \right) \xrightarrow{d} \mathcal{N} . \qquad (7.3)$$

Proof. Let $Y_i = (X_i - \mu)/\sigma$, $i = 1, 2, \ldots$, and denote $\tilde{S}_k = Y_1 + \ldots + Y_k$, $k = 1, 2, \ldots$. Then (7.3) becomes

$$\frac{1}{\sqrt{2n}} \sum_{k=1}^{n} \frac{\tilde{S}_k}{k} \xrightarrow{d} \mathcal{N} .$$

Observe that

$$\sum_{k=1}^{n} \frac{\tilde{S}_k}{k} = \sum_{k=1}^{n} \frac{1}{k} \sum_{i=1}^{k} Y_i = \sum_{i=1}^{n} b_{i,n} Y_i ,$$

where

$$b_{i,n} = \sum_{k=i}^{n} \frac{1}{k} , \quad i = 1, \ldots, n.$$

Define now

$$Z_{i,n} = \frac{b_{i,n}}{\sqrt{2n}} Y_i,$$

and observe that $E(Z_{i,n}) = 0$ and

$$Var(Z_{i,n}) = \frac{b_{i,n}^2}{2n}, \quad i = 1, \ldots, n.$$

Also, since for $k \geq l$,

$$Cov\left(\frac{\tilde{S}_k}{k}, \frac{\tilde{S}_l}{l} \right) = \frac{1}{k},$$

then

$$Var\left(\sum_{i=1}^{n} Z_{i,n}\right) = \frac{1}{2n}Var\left(\sum_{k=1}^{n}\frac{\check{S}_k}{k}\right)\frac{1}{2n}\left(b_{1,n} + 2\sum_{k=2}^{n}\sum_{l=1}^{k-1}\frac{1}{k}\right)$$

$$= \frac{1}{2n}\left(b_{1,n} + 2\sum_{k=2}^{n}\frac{k-1}{k}\right) = 1 - \frac{b_{1,n}}{2n} \to 1$$

as $n \to \infty$. Observe that as a by-product of the above computation we have obtained the identity

$$\sum_{k=1}^{n} b_{i,n}^2 = 2n - b_{1,n}. \tag{7.4}$$

In order to complete the proof we need to check that the Lindeberg condition is satisfied for the triangular array $[Z_{i,n}]$. Take any $\varepsilon > 0$. Then, using (7.4), we get

$$\sum_{i=1}^{n} E(Z_{i,n}^2 I(|Z_{i,n}| > \varepsilon)) = \frac{1}{2n}\sum_{i=1}^{n} b_{i,n}^2 E\left(Y_i^2 I\left(|Y_i| > \frac{\varepsilon\sqrt{2n}}{b_{i,n}}\right)\right)$$

$$\leq \frac{1}{2n}\sum_{i=1}^{n} b_{i,n}^2 E\left(Y_i^2 I\left(|Y_i| > \frac{\varepsilon\sqrt{2n}}{\log(n)}\right)\right) = \frac{a_n}{2n}\sum_{i=1}^{n} b_{i,n}^2 = a_n\left(1 - \frac{b_{1,n}}{2n}\right),$$

where

$$a_n = E\left(Y_i^2 I\left(|Y_i| > \frac{\varepsilon\sqrt{2n}}{\log(n)}\right)\right)$$

does not depend on i. Since $a_n \to 0$ as $n \to \infty$ then the Lindeberg condition holds. □

With the result of the lemma above established, we may now prove the theorem.

Proof of Theorem 7.2.1. The proof relies on the delta-method (or the Taylor series) expansion. In what follows we use only elementary considerations to justify its validity.

Denote $C_k = S_k/(\mu k)$, $k = 1, 2, \ldots$. By the strong law of large numbers it follows that for any $\delta > 0 \ \exists R$ such that $\forall r > R$

$$P(\sup_{k \geq r} |C_k - 1| > \delta) < \delta.$$

Consequently, there exist two sequences $(\delta_m) \downarrow 0$ $(\delta_1 = 1/2)$ and $(R_m) \uparrow \infty$ such that

$$P(\sup_{k \geq R_m} |C_k - 1| > \delta_m) < \delta_m.$$

Take now any real x and any positive integer m. Then

$$P\left(\frac{1}{\gamma\sqrt{2n}}\sum_{k=1}^{n}\log(C_k)\leq x\right)$$

$$= P\left(\frac{1}{\gamma\sqrt{2n}}\sum_{k=1}^{n}\log(C_k)\leq x,\ \sup_{k>R_m}|C_k-1|>\delta_m\right)$$

$$+ P\left(\frac{1}{\gamma\sqrt{2n}}\sum_{k=1}^{n}\log(C_k)\leq x,\ \sup_{k>R_m}|C_k-1|\leq\delta_m\right)$$

$$= A_{m,n}+B_{m,n}$$

and $A_{m,n}<\delta_m$.

To compute $B_{m,n}$ we will expand the logarithm: $\log(1+x)=x-\frac{x^2}{2\,(1+\theta x)^2}$, where $\theta\in(0,1)$ depends on $x\in(-1,1)$. Thus,

$$B_{m,n}$$

$$= P\left(\frac{1}{\gamma\sqrt{2n}}\sum_{k=1}^{R_m}\log(C_k)+\frac{1}{\gamma\sqrt{2n}}\sum_{k=R_m+1}^{n}\log(1+(C_k-1))\leq x,\right.$$

$$\left.\sup_{k>R_m}|C_k-1|\leq\delta_m\right)=$$

and

$$B_{m,n}$$

$$= P\left(\frac{1}{\gamma\sqrt{2n}}\sum_{k=1}^{R_m}\log(C_k)+\frac{1}{\gamma\sqrt{2n}}\sum_{k=R_m+1}^{n}(C_k-1)\right.$$

$$\left.-\frac{1}{\gamma\sqrt{2n}}\sum_{k=R_m+1}^{n}\frac{(C_k-1)^2}{(1+\theta_k(C_k-1))^2}\leq x,\ \sup_{k>R_m}|C_k-1|\leq\delta_m\right)$$

$$= P\left(\frac{1}{\gamma\sqrt{2n}}\sum_{k=1}^{R_m}\log(C_k)+\frac{1}{\gamma\sqrt{2n}}\sum_{k=R_m+1}^{n}(C_k-1)\right.$$

$$\left.-\left[\frac{1}{\gamma\sqrt{2n}}\sum_{k=R_m+1}^{n}\frac{(C_k-1)^2}{(1+\theta_k(C_k-1))^2}\right]I(\sup_{k>R_m}|C_k-1|\leq\delta_m)\leq x\right)$$

$$-P\left(\frac{1}{\gamma\sqrt{2n}}\sum_{k=1}^{R_m}\log(C_k)+\frac{1}{\gamma\sqrt{2n}}\sum_{k=R_m+1}^{n}(C_k-1)\leq x,\ \sup_{k>R_m}|C_k-1|>\delta_m\right)$$

$$= D_{m,n}+F_{m,n}.$$

where θ_k, $k = 1, \ldots, n$ are $(0, 1)$-valued random variables and $F_{m,n} < \delta_m$.
Rewrite $D_{m,n}$ as

$$D_{m,n} = P\left(\frac{1}{\gamma\sqrt{2n}}\sum_{k=1}^{R_m}(\log(C_k) - C_k + 1) + \frac{1}{\gamma\sqrt{2n}}\sum_{k=1}^{n}(C_k - 1)\right.$$
$$\left. - \left[\frac{1}{\gamma\sqrt{2n}}\sum_{k=R_m+1}^{n}\frac{(C_k - 1)^2}{(1 + \theta_k(C_k - 1))^2}\right]I(\sup_{k>R_m}|C_k - 1| < \delta_m) \leq x\right).$$

Observe now that for any fixed m

$$\frac{1}{\gamma\sqrt{2n}}\sum_{k=1}^{R_m}(\log(C_k) - C_k + 1) \xrightarrow{P} 0 \tag{7.5}$$

as $n \to \infty$ (as a matter of fact this sequence converges to zero a.s.).

Note that for $|x| < 1/2$ and any $\theta \in (0, 1)$ it follows that $x^2/(1+\theta x)^2 \leq 4x^2$.
Then for any m

$$\left[\frac{1}{\gamma\sqrt{2n}}\sum_{k=R_m+1}^{n}\frac{(C_k - 1)^2}{(1 + \theta_k(C_k - 1))^2}\right]I(\sup_{k>R_m}|C_k - 1| < \delta_m)$$

$$\leq \frac{4}{\sqrt{2n}}\sum_{k=1}^{n}(C_k - 1)^2 \xrightarrow{P} 0, \tag{7.6}$$

as $n \to \infty$. The above is a consequence of the Markov inequality, since for any $\varepsilon > 0$

$$P\left(\frac{4}{\sqrt{2n}}\sum_{k=1}^{n}(C_k - 1)^2 > \varepsilon\right)$$

$$\leq \frac{4}{\varepsilon\sqrt{2n}}\sum_{k=1}^{n}Var(C_k) = \frac{4}{\varepsilon\sqrt{2n}}\sum_{k=1}^{n}\frac{1}{k} \to 0$$

as $n \to \infty$ for any fixed m.

Since by Lemma 7.2.1 it follows that

$$\frac{1}{\gamma\sqrt{2n}}\sum_{k=1}^{n}(C_k - 1) \xrightarrow{d} N$$

as $n \to \infty$ then by (7.5) and (7.6) we conclude that for any fixed m

$$D_{m,n} \to \Phi(x),$$

where Φ is the standard normal distribution function.

Finally, observe that

$$P\left(\log\left(\frac{\prod_{k=1}^{n} S_k}{n!\mu^n}\right)^{\frac{1}{\gamma\sqrt{n}}} \le x\right) = P\left(\frac{1}{\gamma\sqrt{2n}}\sum_{k=1}^{n}\log(C_k) \le x\right)$$

$$= A_{m,n} + D_{m,n} + F_{m,n}$$

which implies (2) since $A_{m,n} + F_{m,n} < 2\delta_m \to 0$ as $m \to \infty$, uniformly in n.

□

The following example illustrates some applications of the theorem.

Example 7.2.1. In Arnold and Villaseñor (1998) the following identity was proved:

$$T_n = \sum_{k=1}^{n}\log(\sum_{i=1}^{k} X_k) \stackrel{d}{=} -\sum_{i=1}^{n}\tilde{X}_i + nR_n$$

where (X_n) and $(\tilde{X}_n)$ are independent sequences of independent and identically distributed standard exponential random variables and R_n is nth record from an independent sequence of independent and identically distributed Gumbel random variables (R_1 is the first observation). Consequently,

$$\frac{T_n - n\log(n) + n}{\sqrt{2n}} \stackrel{d}{=} \frac{1}{\sqrt{2}}\left(-\frac{\sum_{i=1}^{n}\tilde{X}_i - n}{\sqrt{n}} + \sqrt{n}(R_n - \log(n))\right).$$

Now, in view of the result above, we can apply the argument used by Arnold and Villaseñor (1998) in the reverse order. From Theorem 7.2.1 it follows that the left hand side converges in distribution to the standard normal law and the same holds true by the standard CLT for the first element at the right hand side. Now since the first and the second element at the right-hand side are independent, it follows that $\sqrt{n}(R_n - \log n)$ is asymptotically standard normal, which proves Resnick's limit theorem (Resnick (1973)) for records from the Gumbel distribution.

The following lemma shall be useful in investigating a law of large numbers result corresponding to Theorem 7.2.1.

Lemma 7.2.2 (Toeplitz's lemma). *Let (x_n) be a real sequence such that $x_n \to x < \infty$ as $n \to \infty$. Let $w_{ni} > 0$ be an array of weights for which $\sum_{i=1}^{n} w_{ni} = 1$ for all $n \ge 1$ as well as $\max_{1\le i\le n} w_{ni} \downarrow 0$ as $n \to \infty$. Then $\sum_{i=1}^{n} w_{ni}x_i \to x$.*

Proof. For any $\varepsilon > 0$ and sufficiently large n, N consider

$$\left|\sum_{i=1}^{n} w_{ni}\, x_i - x\right| = \left|\sum_{i=1}^{N} w_{ni}\,(x_i - x) + \sum_{i=N+1}^{n} w_{ni}\,(x_i - x)\right|$$

$$= \sum_{i=1}^{N} w_{ni}\,|x_i - x| + \sum_{i=N+1}^{n} w_{ni}\,|x_i - x| \le 2\varepsilon$$

To argue the last inequality first choose N such that $|x_i - x| \leq \varepsilon$ for all $i > N$ and then choose n such that $w_{ni} \leq \varepsilon/(N \max_{1 \leq i \leq N} |x_i - x|)$. It follows that

$$\sum_{i=1}^{N} w_{ni}|x_i - x| \leq \max|S_i - S| \sum_{i=1}^{N} w_{ni} \leq \varepsilon$$

as well as

$$\sum_{i=N+1}^{n} w_{ni}|x_i - x| \leq \varepsilon \sum_{i=N+1}^{n} w_{ni} \leq \varepsilon$$

where the last inequality follows from $\sum_{i=1}^{n} w_{ni} = 1$. Since ε was arbitrary, the proof is completed. $\square$

Theorem 7.2.2. *Let (X_n) be a sequence of independent and identically distributed, positive and integrable random variables. Denote $\mu = EX_1 > 0$. Then*

$$\left(\frac{\prod_{k=1}^{n} S_k}{n!} \right)^{\frac{1}{n}} \to \mu \quad a.s.$$

Proof. In view of the classical SLLN (cf. Example 3.6.1) we have the convergence $\log(S_k/k) \to \log\mu$ a.s. The desired result follows now directly from Lemma 7.2.2 with $w_{in} = 1/n$ in view of the identity

$$\left(\frac{\prod_{k=1}^{n} S_k}{n!} \right)^{\frac{1}{n}} = \exp\left(\frac{1}{n} \sum_{k=1}^{n} \log(S_k/k) \right).$$

$\square$

7.2.1 Extension to Classical U-statistics

Let us consider a simple extension of Theorem 7.2.1 to the case of non-degenerate U-statistics with fixed kernel function. Hence, in our present notation

$$U_n^{(m)} = U_n = \binom{n}{m}^{-1} \sum_{1 \leq i_1 < \ldots < i_m \leq n} h(X_{i_1}, \ldots, X_{i_m}) \tag{7.7}$$

where now we assume that h is some fixed symmetric real function of m arguments and the X_i's are independent and identically distributed random variables. Define

$$\widehat{U}_n = \frac{m}{n} \sum_{i=1}^{n} (h_1(X_i) - Eh) + Eh.$$

Note that by virtue of the representation (1.16) we may write

$$U_n = \widehat{U}_n + R_n^{(m)} \tag{7.8}$$

where

$$R_n^{(m)} = \binom{n}{m}^{-1} \sum_{1 \le i_1 < \ldots < i_m \le n} H(X_{i_1}, \ldots, X_{i_m}),$$

and

$$H(x_1, \ldots, x_m) = h(x_1, \ldots, x_m) - \sum_{i=1}^{m} (h_1(x_i) - E\,h) - E\,h.$$

Hence the random variable $R_n^{(m)}$ is a U-statistic with the degree of degeneracy one and satisfying

$$Cov(\widehat{U}_n, R_n^{(m)}) = 0$$

as well as, in view of (2.14),

$$n\,Var\,R_n^{(m)} \to 0 \quad \text{as} \quad n \to \infty. \tag{7.9}$$

The result of Theorem 7.2.1 can be extended to U-statistics as follows.

Theorem 7.2.3. *Let U_n be a statistic given by (7.7). Assume $E\,h^2 < \infty$ and $P(h(X_1, \ldots, X_m) > 0) = 1$, as well as $\sigma^2 = Var(h_1(X_1)) \ne 0$. Denote $\mu = E\,h > 0$ and $\gamma = \sigma/\mu > 0$, the coefficient of variation. Then*

$$\left(\prod_{k=m}^{n} \frac{U_k}{\mu} \right)^{\frac{1}{m\gamma\sqrt{n}}} \xrightarrow{d} e^{\sqrt{2}\mathcal{N}},$$

where $\mathcal{N}$ is a standard normal rv.

In order to prove the theorem we shall first consider a more general version of (7.3).

Lemma 7.2.3. *Under the assumptions of Theorem 7.2.3*

$$\frac{1}{m\,\gamma\sqrt{2n}} \sum_{k=m}^{n} \left(\frac{U_k}{\mu} - 1 \right) \xrightarrow{d} \mathcal{N}.$$

Proof. Using the decomposition (7.8) we have

$$\frac{1}{m\,\gamma\sqrt{2n}} \sum_{k=m}^{n} \left(\frac{U_k}{\mu} - 1 \right) = \frac{1}{m\,\gamma\sqrt{2n}} \sum_{k=m}^{n} \left(\frac{\widehat{U}_k}{\mu} - 1 \right) + \frac{1}{m\,\sigma\sqrt{2n}} \sum_{k=m}^{n} R_k.$$

By Lemma 7.2.1, applied to the random variables $m\,h_1(X_i)$ for $i = 1, 2, \ldots$ we have

$$\frac{1}{m\,\gamma\sqrt{2n}} \sum_{k=m}^{n} \left(\frac{\widehat{U}_k}{\mu} - 1 \right)$$

$$= \frac{1}{\gamma\sqrt{2n}} \sum_{k=1}^{n} \left(\frac{\sum_{i=1}^{k} h_1(X_i)}{\mu k} - 1 \right) - \sum_{k=1}^{m-1} \left(\frac{\sum_{i=1}^{k} h_1(X_i)}{\mu k} - 1 \right) \xrightarrow{d} \mathcal{N}.$$

since the second expression converges to zero a.s. as $n \to \infty$. Therefore, in order to prove the lemma it suffices to show

$$\tilde{R}_n = \frac{1}{m\sigma\sqrt{2n}} \sum_{k=m}^{n} R_k \xrightarrow{P} 0 \quad \text{as} \quad n \to \infty.$$

To argue the above, it is, in turn, enough to argue that

$$E \tilde{R}_n^2 \to 0 \quad \text{as} \quad n \to \infty.$$

To this end let us note that, similarly as in (6.26) due to symmetries involved, we have

$$Cov\ (R_l, R_k) = Var\ R_k \quad \text{for} \quad l < k,$$

and thus

$$E \tilde{R}_n^2 = Var\ \tilde{R}_n = \frac{1}{m^2 \sigma^2 2n} Var\ \left(\sum_{k=m}^{n} R_k \right)$$

$$= \frac{1}{m^2 \sigma^2 2n} \left\{ \sum_{k=m}^{n} Var\ R_k + 2 \sum_{m \le l < k \le n} Cov\ (R_k, R_l) \right\}$$

$$= \frac{1}{m^2 \sigma^2 2n} \left\{ \sum_{k=m}^{n} Var\ R_k + 2 \sum_{m \le l < k \le n} Var\ R_k \right\}$$

$$= \frac{1}{m^2 \sigma^2 2n} \left\{ \sum_{k=m}^{n} (1 + 2k - 2m)\,Var\ R_k \right\} \to 0$$

as $n \to \infty$, in view of (7.9) and the Toeplitz's lemma (Lemma 7.2.2) with $w_{ni} = 1/n$. □

With the Lemma 7.2.3 established, the proof of Theorem 7.2.3 follows easily.

Proof of Theorem 7.2.3. In view of the fact that if $E\,|h| < \infty$ then $\binom{n}{m}^{-1} U_n \to Eh = \mu$ a.s. (see, Lemma 3.6.1) and Lemma 7.2.3 above, the argument used in the proof of Theorem 7.2.1 can be virtually repeated with S_k/k replaced now by $\binom{k}{m}^{-1} U_k$ and γ by $m\gamma$. □

Remark 7.2.1. Let us note that in view of the strong law of large numbers for U-statistics we may extend the result of Theorem 7.2.2 on the strong convergence of products of partial sums to the products of U-statistics as follows

$$\left(\prod_{k=m}^{n} U_k \right)^{\frac{1}{n}} \to \mu \quad a.s.$$

if $E\,|h| < \infty$.

7.3 Products of Independent Partial Sums

Having established the limiting result for the products of partial sums of a sequence of random variables, we shall now consider a case when partial sums S_k are mutually independent and have square integrable components. This setup, in spite of being interesting on its own, pertains also to a limit theorem for random determinants of Wishart matrices. The main result is provided in Theorem 7.3.1 and is seen to parallel to large extend, the result of Theorem 7.2.1 discussed earlier. Also its extensions to independent and non-identically distributed case as well as to the case of non-degenerate U-statistics will be discussed. Finally, in the last section after a brief review of some basic facts on Wishart matrices, we explain how the limit theorems for products of increasing sums can be used for studying limiting properties of Wishart determinants.

Our main result of this section is the following analogue of Theorem 7.2.1.

Theorem 7.3.1. *Let* $(X_{k,i})_{i=1,\ldots,k}$; $k = 1, 2, \ldots$ *be a triangular array of independent and identically distributed positive square integrable random variables with finite absolute moment of order* $p > 2$. *Denote* $\mu = E(X_1) > 0$, $\gamma = \sigma/\mu$, *where* $\sigma^2 = Var(X_1)$, *and* $S_k = X_{k,1} + \ldots + X_{k,k}$, $k = 1, 2, \ldots$. *Then as* $n \to \infty$

$$\left(n^{\frac{\gamma^2}{2}} \frac{\prod_{k=1}^{n} S_k}{n! \mu^n}\right)^{\frac{1}{\gamma\sqrt{\log(n)}}} \xrightarrow{d} e^{\mathcal{N}},$$

where $\mathcal{N}$ *is a standard normal rv.*

Before proving the above result we will establish a version of the classical central limit theorem, essentially, for scaled independent and identically distributed random variables, analogously as it was done in the scenario discussed earlier. To this end we will use the CLT for triangular arrays, so again the basic step in the proof will relay on verifying the Lindeberg condition. First we recall an elementary fact about the moments of sums of independent and identically distributed variables (e.g., Lee 1990, p. 22). In the sequel, for notational convenience, we set $C_k = S_k/(\mu k)$.

Lemma 7.3.1 (Burkholder's Inequality). *Let* $p \geq 2$. *Under the assumptions of Theorem 7.3.1 there exists a universal constant* D_p *(i.e., depending on* p *but not on* k) *such that for* $k \geq 1$

$$E\left|(C_k - 1)\right|^p \leq \frac{D_p}{k^{p/2}}.$$

Next we introduce the following result which may be seen as an analogue of Lemma 7.2.1.

Lemma 7.3.2. *Under the assumptions of Theorem 7.3.1, as $n \to \infty$*

$$\frac{1}{\gamma\sqrt{\log(n)}} \sum_{k=1}^{n} (C_k - 1) \xrightarrow{d} \mathcal{N} .$$

Proof. Since

$$Var\left(\sum_{k=1}^{n}\left(\frac{S_k}{\mu k} - 1\right)\right) = \gamma^2 \sum_{k=1}^{n} \frac{1}{k} \to \infty \qquad \text{as } n \to \infty$$

and by Lemma 7.3.1 also

$$\limsup_{n} \sum_{k=1}^{n} E\left|\frac{S_k}{\mu k} - 1\right|^p \le D_p \sum_{k=1}^{\infty} \frac{1}{k^{p/2}} < \infty.$$

Thus

$$\frac{\left(\sum_{k=1}^{n} E\left|\frac{S_k}{\mu k} - 1\right|^p\right)^{2/p}}{Var\left(\sum_{k=1}^{n}\left(\frac{S_k}{\mu k} - 1\right)\right)} \to 0 \qquad \text{as } n \to \infty$$

so the Lyapounov and hence the Lindeberg condition is satisfied.   □

Somewhat unexpectedly, the limiting result in this case of independent sums requires a much more subtle treatment of the elements of the expansion of the logarithmic function. First, it is expanded until the terms of order three, not two as in the proof of Theorem 7.2.1. Second, the second term of the expansion does not converge in probability to zero, also it is not immediate to see that the third term converges to zero. These issues are treated in the auxiliary result below. In particular, the first result giving a kind of weak law of large numbers for squares of standardized sums seems to be of independent interest. In both cases in the proofs a delicate truncation technique is employed.

Lemma 7.3.3. *Under the assumptions of Theorem 7.3.1, as $n \to \infty$*

(i) $\frac{1}{\sqrt{\log(n)}} \sum_{k=1}^{n}\left[(C_k - 1)^2 - \frac{\gamma^2}{k}\right] \xrightarrow{P} 0$

(ii) $\frac{1}{\sqrt{\log(n)}} \sum_{k=1}^{n} |C_k - 1|^3 \xrightarrow{P} 0.$

Proof. For the proof of (i) denote $Z_k = k(C_k-1)^2 - \gamma^2$ and note that $EZ_k = 0$ and by our assumptions and Lemma 7.3.1 there exists $0 < \alpha < 1$ such that $\sup_k E|Z_k|^{1+\alpha} < \infty$. Let a_n be any numeric sequence such that $a_n \to \infty$ but $a_n^{1-\alpha}/\log(n) \to 0$ (e.g., $a_n = \log(n)$ will do). Define $Z_k' = Z_k\, I(|Z_k|/k \le a_n)$ and note that for some universal constant D_α

$$P\left(\sum_{k=1}^{n} Z_k/k \ne \sum_{k=1}^{n} Z_k'/k\right) \le \sum_{k=1}^{n} P\left(|Z_k| > k\, a_n\right) \tag{7.10}$$

$$\le D_\alpha/a_n^{1+\alpha} \to 0 \qquad \text{as } n \to \infty.$$

We now show that for the weighted sum of Z'_k's the weak law of large numbers holds. Indeed, for any $\varepsilon > 0$

$$P\left(\left|\sum_{k=1}^{n}(Z'_k/k - EZ'_k/k)\right| > \varepsilon\sqrt{\log(n)}\right) \leq \frac{1}{\varepsilon^2 \log(n)}\sum_{k=1}^{n}E(Z'_k/k - EZ'_k/k)^2$$

$$\leq \frac{1}{\varepsilon^2 \log(n)}\sum_{k=1}^{n}E|Z'_k/k|^2 \leq \frac{a_n^{1-\alpha}}{\varepsilon^2 \log(n)}\sum_{k=1}^{n}E|Z_k/k|^{1+\alpha} \to 0 \qquad \text{as } n \to \infty.$$

$$(7.11)$$

Finally, note that since $EZ_k = 0$ then

$$\left|\sum_{k=1}^{n}EZ'_k/k\right| = \left|\sum_{k=1}^{n}\frac{1}{k}EZ_k\,I(|Z_k/k| > a_n)\right|$$

$$\leq \frac{1}{a_n^{\alpha}}\sum_{k=1}^{n}E|Z_k/k|^{1+\alpha} \to 0 \qquad \text{as } n \to \infty.$$

$$(7.12)$$

The relations (7.10)–(7.12) imply (i).

In order to show the second assertion we proceed similary denoting this time $W_k = k(C_k - 1)$ and $W'_k = W_k\,I(|W_k| \leq b_n)$ where b_n is any numeric sequence satisfying $b_n \to \infty$ but $b_n^{2\,(2-\alpha)}/\log(n) \to 0$. Note that as before, $\sup_k E|W_k|^{1+\alpha} < \infty$. In this notation, the relation (ii) follows when we replace W_k's with W'_k's in view of the Markov inequality and the fact that

$$\frac{1}{\sqrt{\log(n)}}\sum_{k=1}^{n}E|W'_k/k|^3 \leq \frac{b_n^{2-\alpha}}{\sqrt{\log(n)}}\sum_{k=1}^{n}E|W_k/k|^{1+\alpha} \to 0 \qquad \text{as } n \to \infty.$$

The fact that this also implies (ii) for the W_k's is immediate since

$$P\left(\sum_{k=1}^{n}W_k/k \neq \sum_{k=1}^{n}W'_k/k\right) \leq \sum_{k=1}^{n}P\left(|W_k| > k\,b_n\right) \qquad (7.13)$$

$$\leq \sum_{k=1}^{n}E|W_k/(k\,b_n)|^{1+\alpha} \leq D_{\alpha}/b_n^{1+\alpha} \to 0 \quad \text{as } n \to \infty. \qquad (7.14)$$

$$\square$$

Finally, we are in position to prove the main result of this section.

Proof of Theorem 7.3.1. We first note that $C_k = S_k/(\mu k)$ converges almost surely to one. Indeed, for any $\delta > 0$ we have

$$P(\sup_{k \geq r}|C_k - 1| > \delta) \leq \sum_{k=r}^{\infty}P(|C_k - 1| > \delta) \leq \frac{1}{\delta}\sum_{k=r}^{\infty}E|C_k - 1|^p \to 0 \quad \text{as } r \to \infty.$$

Consequently, there exist two sequences $(\delta_m) \downarrow 0$ $(\delta_1 = 1/2)$ and $(R_m) \uparrow \infty$ such that

$$P(\sup_{k \geq R_m} |C_k - 1| > \delta_m) < \delta_m.$$

Take now any real x and any m. Then

$$P\left(\frac{1}{\gamma\sqrt{\log(n)}} \sum_{k=1}^{n} \left(\log(C_k) + \frac{\gamma^2}{2k} \right) \leq x \right)$$

$$= P\left(\frac{1}{\gamma\sqrt{\log(n)}} \sum_{k=1}^{n} \left(\log(C_k) + \frac{\gamma^2}{2k} \right) \leq x \,,\, \sup_{k > R_m} |C_k - 1| > \delta_m \right)$$

$$+ P\left(\frac{1}{\gamma\sqrt{\log(n)}} \sum_{k=1}^{n} \left(\log(C_k) + \frac{\gamma^2}{2k} \right) \leq x \,,\, \sup_{k > R_m} |C_k - 1| \leq \delta_m \right)$$

$$= A_{m,n} + B_{m,n}$$

and $A_{m,n} < \delta_m$.

To compute $B_{m,n}$ we will expand the logarithm: $\log(1 + x) = x - \frac{x^2}{2} + \frac{x^3}{3(1+\theta x)^3}$, where $\theta \in (0,1)$ depends on $x \in (-1,1)$. Thus,

$$B_{m,n}$$

$$= P\left(\frac{1}{\gamma\sqrt{\log(n)}} \sum_{k=1}^{R_m} \left(\log(C_k) + \frac{\gamma^2}{2k} \right) \right.$$

$$+ \frac{1}{\gamma\sqrt{\log(n)}} \sum_{k=R_m+1}^{n} \left(\log(1 + (C_k - 1)) + \frac{\gamma^2}{2k} \right) \leq x, \left. \sup_{k > R_m} |C_k - 1| \leq \delta_m \right)$$

$$= P\left(\frac{1}{\gamma\sqrt{\log(n)}} \sum_{k=1}^{R_m} \left(\log(C_k) + \frac{\gamma^2}{2k} \right) + \frac{1}{\gamma\sqrt{\log(n)}} \sum_{k=R_m+1}^{n} (C_k - 1) \right.$$

$$- \frac{1}{2\gamma\sqrt{\log(n)}} \sum_{k=R_m+1}^{n} \left[(C_k - 1)^2 - \frac{\gamma^2}{k} \right]$$

$$+ \frac{1}{3\gamma\sqrt{\log(n)}} \sum_{k=R_m+1}^{n} \frac{(C_k - 1)^3}{(1 + \theta_k(C_k - 1))^3} \leq x, \left. \sup_{k > R_m} |C_k - 1| \leq \delta_m \right)$$

$$= P\left(\frac{1}{\gamma\sqrt{\log(n)}} \sum_{k=1}^{R_m} \left(\log(C_k) + \frac{\gamma^2}{2k} \right) + \frac{1}{\gamma\sqrt{\log(n)}} \sum_{k=R_m+1}^{n} (C_k - 1) \right.$$

$$- \frac{1}{2\gamma\sqrt{\log(n)}} \sum_{k=R_m+1}^{n} \left[(C_k - 1)^2 - \frac{\gamma^2}{k} \right]$$

$$\left. + \frac{1}{3\gamma\sqrt{\log(n)}} \left[\sum_{k=R_m+1}^{n} \frac{(C_k - 1)^3}{(1 + \theta_k(C_k - 1))^3} \right] I \left(\sup_{k > R_m} |C_k - 1| \leq \delta_m \right) \leq x \right)$$

$$- P \left(\frac{1}{\gamma \sqrt{\log(n)}} \sum_{k=1}^{R_m} \left(\log(C_k) + \frac{\gamma^2}{2k} \right) + \frac{1}{\gamma \sqrt{\log(n)}} \sum_{k=R_m+1}^{n} (C_k - 1) + \right.$$

$$\left. - \frac{1}{2\gamma \sqrt{\log(n)}} \sum_{k=R_m+1}^{n} \left[(C_k - 1)^2 - \frac{\gamma^2}{k} \right] \le x, \ \sup_{k > R_m} |C_k - 1| > \delta_m \right)$$

$$= D_{m,n} + F_{m,n}.$$

where θ_k, $k = 1, \dots, n$ are $(0, 1)$-valued random variables and $F_{m,n} < \delta_m$.
Rewrite now $D_{m,n}$ as

$$D_{m,n} = P \left(\frac{1}{\gamma \sqrt{\log(n)}} \sum_{k=1}^{R_m} \left(\log(C_k) - C_k + 1 + \frac{(C_k - 1)^2}{2} - \frac{\gamma^2}{2k} \right) \right.$$

$$+ \frac{1}{\gamma \sqrt{\log(n)}} \sum_{k=1}^{n} (C_k - 1) - \frac{1}{2\gamma \sqrt{\log(n)}} \sum_{k=1}^{n} \left[(C_k - 1)^2 - \frac{\gamma^2}{k} \right]$$

$$+ \frac{1}{3\gamma \sqrt{\log(n)}} \left[\sum_{k=R_m+1}^{n} \frac{(C_k - 1)^3}{(1 + \theta_k(C_k - 1))^3} \right]$$

$$\left. \times I \left(\sup_{k > R_m} |C_k - 1| < \delta_m \right) \le x \right).$$

Observe that for any fixed m

$$\frac{1}{\gamma \sqrt{2n}} \sum_{k=1}^{R_m} \left(\log(C_k) - C_k + 1 + \frac{(C_k - 1)^2}{2} - \frac{\gamma^2}{2k} \right) \overset{P}{\to} 0 \quad \text{as } n \to \infty \quad (7.15)$$

(as a matter of fact, this sequence converges to zero a.s.).
Invoking Lemma 7.3.3(i) we see

$$P \left(\frac{1}{2\gamma \sqrt{\log(n)}} \sum_{k=1}^{n} \left[(C_k - 1)^2 - \frac{\gamma^2}{k} \right] > \varepsilon \right) \to 0.$$

Note that for $|x| < 1/2$ and any $\theta \in (0, 1)$ it follows that $|x|^3/|1 + \theta x|^3 \le 8|x|^3$.
Thus for any m

$$\frac{1/3}{\gamma \sqrt{\log(n)}} \left[\sum_{k=R_m+1}^{n} \frac{|C_k - 1|^3}{|1 + \theta_k(C_k - 1)|^3} \right] I \left(\sup_{k > R_m} |C_k - 1| < \delta_m \right)$$

$$\le \frac{8/3}{\gamma \sqrt{\log(n)}} \sum_{k=1}^{n} |C_k - 1|^3 \overset{P}{\to} 0, \qquad (7.16)$$

as $n \to \infty$ by Lemma 7.3.3(ii).

Since, on the other hand, by Lemma 7.3.1 it follows that

$$\frac{1}{\gamma\sqrt{\log(n)}} \sum_{k=1}^{n} (C_k - 1) \xrightarrow{d} \mathcal{N}$$

as $n \to \infty$ then by (7.15) and (7.16) we conclude that for any fixed m

$$D_{m,n} \to \Phi(x),$$

where Φ is the standard normal distribution function.

Finally, observe that

$$P\left(\log\left(n^{\frac{\gamma^2}{2}} \frac{\prod_{k=1}^{n} S_k}{n! \mu^n}\right)^{\frac{1}{\gamma\sqrt{\log(n)}}} \le x\right)$$

$$= P\left(\frac{1}{\gamma\sqrt{\log(n)}}\left(\sum_{k=1}^{n} \log(C_k) + \frac{\gamma^2}{2}\log(n)\right) \le x\right)$$

$$= A_{m,n} + D_{m,n} + F_{m,n}$$

which implies the assertion of Theorem 7.3.1, since $A_{m,n} + F_{m,n} < 2\delta_m \to 0$ as $m \to \infty$, uniformly in n and

$$\frac{\log(n) - \sum_{k=1}^{n}\frac{1}{k}}{\sqrt{\log(n)}} \to 0 \quad \text{as } n \to \infty. \qquad \square$$

Similarly as in the case of sequences, it is perhaps worth to notice at this point that the law of large numbers for our present case is again particularly simple. Indeed, as soon as we have $S_k/k \to \mu$ a.s. (in particular, under the assumptions of Theorem 7.3.1) then by the property of the geometric mean (cf. Theorem 7.2.2) it follows directly that as $n \to \infty$

$$\left(\frac{\prod_{k=1}^{n} S_k}{n!}\right)^{\frac{1}{n}} \to \mu \quad a.s. \tag{7.17}$$

Let us state it for the record as

Theorem 7.3.2. *Let* $(X_{k,i})_{i=1,\dots,k}$, $k = 1, 2, \dots$ *be a triangular array of independent and identically distributed, positive and integrable random variables. Denote* $\mu = EX_{1,1} > 0$. *Then* (7.17) *holds.* $\qquad \square$

7.3.1 Extensions

The following extension of Theorem 7.3.1 covering the non-all-independent and identically distributed setting is rather straightforward.

Theorem 7.3.3. *Let $(X_{k,i})_{i=1,\ldots,k}$; $k = 1, 2, \ldots$ be a triangular array of independent and row-wise identically distributed, positive random variables with finite absolute moment of order $p > 2$. Denote, as before, $S_k = X_{k,1} + \ldots + X_{k,k}$, with $\mu_k = E(X_{k,1}) > 0$, $\sigma_k^2 = Var(X_{k,1})$, and $\gamma_k = \sigma_k/\mu_k$. Let $c_n^2 = \sum_{k=1}^{n} \gamma_k^2/k$. If*

(i) $c_n \to \infty$ as $n \to \infty$

(ii) $\sum_{k=1}^{\infty} E \left| \frac{S_k - k\mu_k}{k\mu_k} \right|^p < \infty$,

then as $n \to \infty$

$$\left(e^{\frac{c_n^2}{2}} \frac{\prod_{k=1}^{n} S_k/\mu_k}{n!} \right)^{\frac{1}{c_n}} \xrightarrow{d} e^{\mathcal{N}},$$

where $\mathcal{N}$ is a standard normal rv.

The extension of the result of Theorem 7.3.1 to U-statistics may be done similarly as in Section 7.2.1. Following that section notation we thus have

Theorem 7.3.4. *Let $(X_{k,i})_{i=1,\ldots,k}$; $k = m, m+1, \ldots$ be a triangular array of independent and identically distributed random variables. Let U_k be a statistic given by (7.7) based on $X_{k,1}, \ldots, X_{k,k}$. Assume $E |h|^p < \infty$ for some $p > 2$ and $P(h(X_1, \ldots, X_m) > 0) = 1$, as well as $\sigma^2 = m^2 Var(h_1(X_1)) \neq 0$. Denote $\mu = E h > 0$ and let $\gamma = \sigma/\mu > 0$ be the coefficient of variation. Then, as $n \to \infty$*

$$\left(n^{\frac{\gamma^2}{2}} \frac{\prod_{k=1}^{n} U_k}{\mu^n} \right)^{\frac{1}{\gamma\sqrt{\log(n)}}} \xrightarrow{d} e^{\mathcal{N}},$$

where $\mathcal{N}$ is a standard normal rv.

Proof. Set now $C_k = U_k/\mu$ and let γ be defined as above. Retaining the notation of the previous section with these modifications we find that the result of Lemma 7.3.1 still holds true (cf. e.g., Lee 1990, p. 21). Regarding the extension of the conclusion of Lemma 7.3.2 set $z_n = \gamma(\log(n))^{-1/2}$ and note that by (7.8)

$$z_n \sum_{k=m}^{n} (C_k - 1) = z_n \sum_{k=m}^{n} \left(\frac{\widehat{U}_k}{\mu} - 1 + \frac{R_n}{\mu} \right) = z_n \sum_{k=m}^{n} \left(\frac{\widehat{U}_k}{\mu} - 1 \right) + z_n \sum_{k=m}^{n} \frac{R_n}{\mu}.$$

The first sum in the latest expression above is asymptotically standard normal in view of the result of Lemma 7.3.2 of Section 2 and the second one vanishes asymptotically in probability since by (7.9)

$$z_n^2 Var \left(\sum_{k=1}^{n} \frac{R_n}{\mu} \right) = z_n^2 \sum_{k=1}^{n} \left(\frac{Var R_n}{\mu^2} \right) \to 0 \quad \text{as } n \to \infty.$$

Hence, the result of Lemma 7.3.2 remains valid for U-statistics and a similar argument can be invoked to argue that the results of Lemma 7.3.3 are true for

U-statistics as well. Finally, in view of the SLLN for U-statistics (see, Lemma 3.6.1) which implies that under our assumptions $C_k \to 1$ as $k \to \infty$, we may virtually repeat the expansion argument used in the proof of Theorem 7.3.1 to obtain the required assertion. $\square$

7.4 Asymptotics for Wishart Determinants

In this section we are concerned with random matrices of the form $\sum Y_i Y_i^T$, where Y_i's are independent identically distributed Gaussian (column) vectors. Such matrices, called Wishart matrices $W_n(n, \Sigma_n)$, $n = 1, 2, \ldots$, are of special importance in multivariate statistical analysis (in particular, the distribution of the sample covariance matrix for observations from multivariate normal distribution is Wishart) and were widely investigated since their introduction in late twenties of the last century. More recently asymptotic properties of Wishart matrices has been studied. The investigations in this direction led to the discovery of the famous Marchenko-Pastur law as a weak limit for empirical measures for eigenvalues of Wishart matrices of increasing dimensions. Also, a related issue of limiting behaviour of determinants of such matrices was investigated in the literature. The methods which were used to derive the limiting laws were rather involved and the derivations extremely complicated. However, the determinant of the Wishart matrix is a product of independent increasing sums. Therefore it fits well the topic of the present chapter. Consequently, the derivation of the limiting law for the Wishart determinants which is offered herein is a simple consequence of the asymptotic result for the product of independent sums as given in Theorem 7.3.1.

First, some basic facts about the classical Wishart distribution will be recalled.

Definition 7.4.1 (Wishart distribution). *Let $Y_1, \ldots, Y_n$ be independent and identically distributed d-dimensional Gaussian zero-mean random vectors with a positive definite covariance matrix Σ. The $d \times d$ dimensional random matrix $\mathbf{A} = \sum_{i=1}^n Y_i Y_i^T$ is said to have the classical Wishart distribution $W_d(n, \Sigma)$.*

If $n \geq d$ then the distribution of $\mathbf{A}$ is concentrated on the open cone of $d \times d$ positive definite symmetric matrices $\mathcal{V}_d^+$ and its density with respect to the appropriate Lebesgue measure is

$$f(x) = \frac{\det(x)^{\frac{n-d-1}{2}} \exp[-\frac{1}{2}(\sigma^{-1}, x)]}{2^{\frac{nd}{2}} \det(\Sigma)^{\frac{n}{2}} \Gamma_d\left(\frac{n}{2}\right)} I_{\mathcal{V}_d^+}(x).$$

If $n < d$ then the Wishart measure is singular with respect to the Lebesgue measure and is concentrated on the boundary of the cone $\mathcal{V}_d^+$.

Let $\mathbf{A} \sim W_d(n, \Sigma)$ be decomposed into blocks according to the dimensions p and q, $p + q = d$

$$\mathbf{A} = \begin{pmatrix} \mathbf{A}_1 & \mathbf{A}_{12} \\ \mathbf{A}_{21} & \mathbf{A}_2 \end{pmatrix},$$

such that $\mathbf{A_1}$ is a $p \times p$, $\mathbf{A}_{12} = \mathbf{A}_{21}^T$ is a $p \times q$ and $\mathbf{A}_2$ is a $q \times q$ matrix. Similarly we can decompose

$$\Sigma = \begin{pmatrix} \Sigma_1 & \Sigma_{12} \\ \Sigma_{21} & \Sigma_2 \end{pmatrix}.$$

It is well known that:

1. $\mathbf{A}_1 \sim W_p(n, \Sigma_1)$;
2. $\mathbf{A}_{2\cdot1} = \mathbf{A}_2 - \mathbf{A}_{21}\mathbf{A}_1^{-1}\mathbf{A}_{12} \sim W_q(n-p, \Sigma_{2\cdot1})$ with $\Sigma_{2\cdot1} = \Sigma_2 - \Sigma_{21}\Sigma_1^{-1}\Sigma_{12}$;
3. the pair $(\mathbf{A}_1, \mathbf{A}_{12})$ and $\mathbf{A}_{2\cdot1}$ are independent;
4. $\det(\mathbf{A}) = \det(\mathbf{A}_1)\det(\mathbf{A}_{2\cdot1})$.

Now we decompose $\mathbf{A} = [a_{ij}]$ step by step. First $\mathbf{A} = \mathbf{A}_{(1...d)}$ into blocks: $\mathbf{A}_{(1...d-1)}$ of dimensions $(d-1) \times (d-1)$ and $\mathbf{A}_d = a_{dd}$ of dimensions 1×1, then $\mathbf{A}_{(1...d-1)}$ into blocks $\mathbf{A}_{(1...d-2)}$ of dimensions $(d-2) \times (d-2)$ and $\mathbf{A}_{d-1} = a_{d-1,d-1}$ of dimensions 1×1, ending up with $\mathbf{A}_{(12)}$ decomposed into $\mathbf{A}_1 = a_{11}$ and $\mathbf{A}_2 = a_{22}$ both of dimensions 1×1. By properties (1-4) above we have the following multiplicative representation for the determinant of $\mathbf{A}$

$$\det(\mathbf{A}) = \mathbf{A}_{d\cdot(1...d-1)}\mathbf{A}_{d-1\cdot(1...d-2)} \cdot \ldots \cdot \mathbf{A}_{2\cdot1}\mathbf{A}_1,$$

where the factors are independent gamma variables:

$$Y_{n+1-k} = \mathbf{A}_{k\cdot1...k-1} \sim G\left(\frac{n+1-k}{2}, \frac{1}{2\Sigma_{k\cdot(1...k-1)}}\right), \quad k = 1,\ldots,d$$

understanding that $Y_1 = \mathbf{A}_1$. Here the gamma distribution $G(a, p)$ is defined through its density of the form $f(x) \propto x^{p-1}e^{-ax}I_{(0,\infty)}(x)$ for $a, p > 0$. Thus for a triangular array of independent and identically distributed chi-square variables with one degree of freedom X_{kj}, $j = 1,\ldots,k$, $k = 1, 2,\ldots, d$, we have

$$Y_k \stackrel{d}{=} c_{kn} \sum_{l=1}^{k} X_{kl},$$

where $c_{kn} = \Sigma_{n+1-k\cdot(1...n-k)}$, $k = n - d + 1,\ldots, n$.

Taking all what was said above into account and using Theorem 7.3.1 we can obtain the following asymptotic result for determinants of Wishart matrices.

Theorem 7.4.1. *Let* $\mathbf{A}_n \sim W_n(n, \Sigma_n)$, $n = 1, 2,\ldots$. *Then*

$$\left(\frac{\det(\mathbf{A}_n)}{\det(\Sigma_n)(n-1)!}\right)^{\frac{1}{\sqrt{2\log(n)}}} \stackrel{d}{\to} e^{\mathcal{N}}. \tag{7.18}$$

Proof. Note that putting $d = n$ in the considerations prior to the formulation of Theorem 7.4.1, we have

$$\frac{\det(\mathbf{A}_n)}{\prod_{k=1}^{n} c_{kn}} \stackrel{d}{=} \prod_{k=1}^{n} \left(\sum_{l=1}^{n} X_{kl} \right),$$

where (X_{kl}) are independent and identically distributed $\chi^2(1)$ random variables. Thus in the notation of Theorem 7.3.1 we have $\mu = 1$ and $\gamma^2 = 2$. Moreover $\prod_{k=1}^{n} c_{kn} = \det(\Sigma_n)$. Now the result follows directly from Theorem 7.3.1.   □

7.5 Bibliographic Details

The asymptotic behavior of a product of partial sums of a sequence of independent and identically distributed, positive random variables has been discussed in several relatively recent papers (see, e.g., Qi (2003) for a brief review). In particular, the results in Theorems 7.2.1 are from Rempała and Wesołowski (2002a) and Rempała and Wesołowski (2005b). Further results extending the ones discussed in this chapter to the case when the underling distribution is in the domain of attraction of a stable law with index from the interval [1,2] are obtained in Qi (2003) and Lu and Qi (2004). Another extension of these results is given in Gonchigdanzan (2005) by means of the so-called 'almost sure' versions of the limiting results of Theorems 7.2.1 and 7.2.3.

The standard references on multivariate analysis and in particular on Wishart distributions are Anderson (1984) or Muirhead (1982). The theory of random determinants, including limit theorems covering also non-Wishart case are reviewed for instance in Girko (1997, 1990) and Dembo (1989).

References

Abramowitz, M. and Stegun, I. A. (1964). *Handbook of mathematical functions with formulas, graphs, and mathematical tables*, volume 55 of *National Bureau of Standards Applied Mathematics Series*. For sale by the Superintendent of Documents, U.S. Government Printing Office, Washington, D.C.

Aldous, D. (1985). Exchangability and related topics. In Hennequin, P., editor, *École d'Été de Probabilités de Saint-Flour XIII - 1983*, volume 1117 of *Lecture Notes in Mathematics*, pages 1–198. Springer-Verlag.

Anderson, T. W. (1984). *An introduction to multivariate statistical analysis*. Wiley Series in Probability and Mathematical Statistics: Probability and Mathematical Statistics. John Wiley & Sons Inc., New York, second edition.

Avram, F. and Taqqu, M. S. (1986). Symmetric polynomials of random variables attracted to an infinitely divisible law. *Probab. Theory Relat. Fields*, 71(4):491–500.

Bagchi, S. (1994). Optimality and construction of some rectangular designs. *Metrika*, 41(1):29–41.

Bagchi, S. and Cheng, C.-S. (1993). Some optimal designs of block size two. *J. Statist. Plann. Inference*, 37(2):245–253.

Bapat, R. B. (1990). Permanents in probability and statistics. *Linear Algebra and Applications*, 127:3–25.

Billingsley, P. (1995). *Probability and measure*. John Wiley & Sons Inc., New York, third edition. A Wiley-Interscience Publication.

Billingsley, P. (1999). *Convergence of probability measures*. John Wiley & Sons Inc., New York, second edition. A Wiley-Interscience Publication.

Binet, J. P. M. (1812). Memoire sur un systeme de formules analytiques, et leur application à des considérations géometriques.[memoirs on the system of analytic formulae and their application to geometric considerations.]. *Journal de L' École Politechnique*, 9.

Blom, G. (1976). Some properties of incomplete U-statistics. *Biometrika*, 63(3):573–580.

Borovskikh, Y. V. and Korolyuk, V. S. (1994). Random permanents and symmetric statistics. *Acta Appl. Math.*, 36(3):227–288.

Bose, R. C., Shrikhande, S. S., and Parker, E. T. (1960). Further results on the construction of mutually orthogonal Latin squares and the falsity of Euler's conjecture. *Canad. J. Math.*, 12:189–203.

Brown, B. M. and Kildea, D. G. (1978). Reduced U-statistics and the Hodges-Lehmann estimator. *Ann. Statist.*, 6(4):828–835.

Brualdi, R. A. and Ryser, H. J. (1991). *Combinatorial matrix theory*, volume 39 of *Encyclopedia of Mathematics and its Applications*. Cambridge University Press, Cambridge.

Bryc, W. (1995). *The normal distribution*, volume 100 of *Lecture Notes in Statistics*. Springer-Verlag, New York. Characterizations with applications.

Cauchy, A. (1812). Mémoire sur les fonctions qui ne peuvent obtenir que deux valeurs égales et de signes contraires per suite des transpositions opérées entre les variables quélles renfermen. *Journal de L' École Politechnique*, 10–17:29–112.

Cheng, C.-S. and Bailey, R. A. (1991). Optimality of some two-associate-class partially balanced incomplete-block designs. *Ann. Statist.*, 19(3):1667–1671.

Chow, Y. S. and Teicher, H. (1978). *Probability theory*. Springer-Verlag, New York. Independence, interchangeability, martingales.

de la Peña, V. H. and Giné, E. (1999). *Decoupling*. Probability and its Applications (New York). Springer-Verlag, New York. From dependence to independence, Randomly stopped processes. U-statistics and processes. Martingales and beyond.

Dembo, A. (1989). On random determinants. *Quart. Appl. Math.*, 47(2):185–195.

Dynkin, E. B. and Mandelbaum, A. (1983). Symmetric statistics, Poisson point processes, and multiple Wiener integrals. *Ann. Statist.*, 11(3):739–745.

Enqvist, E. (1985). A note on incomplete U-statistics for stationary absolutely regular processes. In *Contributions to probability and statistics in honour of Gunnar Blom*, pages 97–103. University of Lund, Lund, Germany.

Erdős, P. and Rényi, A. (1959). On random graphs. I. *Publ. Math. Debrecen*, 6:290–297.

Ethier, S. N. and Kurtz, T. G. (1986). *Markov processes*. Wiley Series in Probability and Mathematical Statistics: Probability and Mathematical Statistics. John Wiley & Sons Inc., New York. Characterization and convergence.

Feller, W. (1968). *An introduction to probability theory and its applications. Vol. I.* Third edition. John Wiley & Sons Inc., New York.

Frees, E. W. (1989). Infinite order U-statistics. *Scand. J. Statist.*, 16(1):29–45.

Fyodorov, Y. V. (2006). On permanental polynomials of certain random matrices. *Int. Math. Res. Not.*, pages Art. ID 61570, 37.

Giné, E., Kwapień, S., Latała, R., and Zinn, J. (2001). The LIL for canonical U-statistics of order 2. *Ann. Probab.*, 29(1):520–557.

Giné, E. and Zinn, J. (1994). A remark on convergence in distribution of U-statistics. *Ann. Probab.*, 22(1):117–125.

Girko, V. L. (1990). *Theory of random determinants*, volume 45 of *Mathematics and its Applications (Soviet Series)*. Kluwer Academic Publishers Group, Dordrecht. Translated from the Russian.

Girko, V. L. (1997). A refinement of the central limit theorem for random determinants. *Teor. Veroyatnost. i Primenen.*, 42(1):63–73.

Gonchigdanzan, K. (2005). A note on the almost sure limit theorem for U-statistic. *Period. Math. Hungar.*, 50(1-2):149–153.

Guillaume, J. and Latapy, M. (2004). Bipartite structure of all complex networks. *Information Processing Letters*, 95(5):215–221.

Halász, G. and Székely, G. J. (1976). On the elementary symmetric polynomials of independent random variables. *Acta Math. Acad. Sci. Hungar.*, 28(3-4):397–400.

Halmos, P. R. (1946). The theory of unbiased estimation. *Ann. Math. Statistics*, 17:34–43.

Herrndorf, N. (1986). An invariance principle for reduced U-statistics. *Metrika*, 33(3-4):179–188.

Hoeffding, W. (1948). A class of statistics with asymptotically normal distribution. *Ann. Math. Statistics*, 19:293–325.

Hoeffding, W. (1961). Strong law of large numbers for U. Technical Report 302, University of North Carolina, Department of Statistics.

Horák, P., Rosa, A., and Sirán, J. (1997). Maximal orthogonal Latin rectangles. *Ars Combin.*, 47:129–145.

Itô, K. (1951). Multiple Wiener integral. *J. Math. Soc. Japan*, 3:157–169.

Jakubowski, A., Mémin, J., and Pagès, G. (1989). Convergence en loi des suites d'intégrales stochastiques sur l'espace $\mathbf{D}^1$ de Skorokhod. *Probab. Theory Related Fields*, 81(1):111–137.

Janson, S. (1984). The asymptotic distributions of incomplete U-statistics. *Z. Wahrsch. Verw. Gebiete*, 66(4):495–505.

Janson, S. (1994). The numbers of spanning trees, Hamilton cycles and perfect matchings in a random graph. *Combin. Probab. Comput.*, 3(1):97–126.

Jerrum, M., Sinclair, A., and Vigoda, E. (2004). A polynomial-time approximation algorithm for the permanent of a matrix with nonnegative entries. *Journal of the ACM*, 51(4):671–697.

Kaneva, E. Y. and Korolyuk, V. S. (1996). Random permanents of mixed sample matrices. *Ukrain. Mat. Zh.*, 48(1):44–49.

Koroljuk, V. S. and Borovskich, Y. V. (1994). *Theory of U-statistics*, volume 273 of *Mathematics and its Applications*. Kluwer Academic Publishers Group, Dordrecht. Translated from the 1989 Russian original by P. V. Malyshev and D. V. Malyshev and revised by the authors.

Korolyuk, V. S. and Borovskikh, Y. V. (1991). Random permanents and symmetric statistics. *Akad. Nauk Ukrain. SSR Inst. Mat. Preprint*, 1991(12):61.

Korolyuk, V. S. and Borovskikh, Y. V. (1992). Random permanents and symmetric statistics. In *Probability theory and mathematical statistics (Kiev, 1991)*, pages 176–187. World Sci. Publishing, River Edge, NJ.

Korolyuk, V. S. and Borovskikh, Y. V. (1995). Normal approximation of random permanents. *Ukrain. Mat. Zh.*, 47(7):922–927.

Kuhn, H. W. (1955). The hungarian method for the assignment problem. *Naval Research Logistic Quarterly*, 2:83–97.

Kuo, H.-H. (2006). *Introduction to stochastic integration*. Universitext. Springer, New York.

Kurtz, T. G. and Protter, P. (1991). Weak limit theorems for stochastic integrals and stochastic differential equations. *Ann. Probab.*, 19(3):1035–1070.

Latała, R. and Zinn, J. (2000). Necessary and sufficient conditions for the strong law of large numbers for U-statistics. *Ann. Probab.*, 28(4):1908–1924.

Lee, A. J. (1982). On incomplete U-statistics having minimum variance. *Austral. J. Statist.*, 24(3):275–282.

Lee, A. J. (1990). *U-statistics.Theory and practice*, volume 110 of *Statistics: Textbooks and Monographs*. Marcel Dekker Inc., New York.

Lu, X. and Qi, Y. (2004). A note on asymptotic distribution of products of sums. *Statist. Probab. Lett.*, 68(4):407–413.

Major, P. (1981). *Multiple Wiener-Itô integrals*, volume 849 of *Lecture Notes in Mathematics*. Springer, Berlin. With applications to limit theorems.

Minc, H. (1978). *Permanents*. Addison-Wesley Publishing Co., Reading, Mass. With a foreword by Marvin Marcus, Encyclopedia of Mathematics and its Applications, Vol. 6.

Móri, T. F. and Székely, G. J. (1982). Asymptotic behaviour of symmetric polynomial statistics. *Ann. Probab.*, 10(1):124–131.

Muirhead, R. J. (1982). *Aspects of multivariate statistical theory*. John Wiley & Sons Inc., New York. Wiley Series in Probability and Mathematical Statistics.

Munkres, J. (1957). Algorithms for the assignment and transportation problems. *Journal of the Society of Industrial and Applied Mathematics*, 5(1):32–38.

Murata, T. (1989). Petri nets: properties analysis and applications. *Proc. of IEEE*, 77(4):541–580.

Newman, M., Watts, D. J., and Strogatz, S. H. (2002). Random graph models of social networks. *Proc. Nat. Acad. of Sc.*, 99:2566–2572.

Nowicki, K. and Wierman, J. C. (1988). Subgraph counts in random graphs using incomplete U-statistics methods. In *Proceedings of the First Japan Conference on Graph Theory and Applications (Hakone, 1986)*, volume 72, pages 299–310.

Peterson, J. L. (1981). *Petri net theory and the modeling of systems*. Prentice-Hall Inc., Englewood Cliffs, N.J.

Politis, D. N. and Romano, J. P. (1994). Large sample confidence regions based on subsamples under minimal assumptions. *Ann. Statist.*, 22(4):2031–2050.

Qi, Y. (2003). Limit distributions for products of sums. *Statist. Probab. Lett.*, 62(1):93–100.

Raghavarao, D. (1988). *Constructions and combinatorial problems in design of experiments.* Dover Publications Inc., New York. Corrected reprint of the 1971 original.

Rempała, G. (2001). The martingale decomposition and approximation theorems for a generalized random permanent function. *Demonstratio Math.*, 34(2):431–446.

Rempała, G. (2004). Asymptotic factorial powers expansions for binomial and negative binomial reciprocals. *Proc. Amer. Math. Soc.*, 132(1):261–272 (electronic).

Rempala, G. and Gupta, A. (2000). Almost sure behavior of elementary symmetric polynomials. *Random Oper. Stochastic Equations*, 8(1):39–50.

Rempała, G. and Srivastav, S. (2004). Minimum variance rectangular designs for U-statistics. *J. Statist. Plann. Inference*, 120(1-2):103–118.

Rempała, G. and Székely, G. (1998). On estimation with elementary symmetric polynomials. *Random Oper. Stochastic Equations*, 6(1):77–88.

Rempała, G. and Wesołowski, J. (1999). Asymptotic behavior of random permanents. *Statist. Probab. Lett.*, 45:149–158.

Rempała, G. and Wesołowski, J. (2002a). Asymptotics for products of sums and U-statistics. *Electron. Comm. Probab.*, 7:47–54 (electronic).

Rempała, G. and Wesołowski, J. (2002b). Central limit theorems for random permanents with correlation structure. *J. Theoret. Probab.*, 15(1):63–76.

Rempała, G. and Wesołowski, J. (2002c). Strong laws of large numbers for random permanents. *Probab. Math. Statist.*, 22(2):201–209.

Rempała, G. and Wesołowski, J. (2003). Incomplete U-statistics of permanent design. *J. Nonparametr. Stat.*, 15(2):221–236.

Rempała, G. and Wesołowski, J. (2004). Limit theorems for random permanents with exchangeable structure. *Journal of Multivariate Analysis.*

Rempała, G. and Wesołowski, J. (2005a). Approximation theorems for random permanents and associated stochastic processes. *Probab. Theory Related Fields*, 131(3):442–458.

Rempała, G. and Wesołowski, J. (2005b). Asymptotics for products of independent sums with an application to Wishart determinants. *Statist. Probab. Lett.*, 74(2):129–138.

Rempała, G. and Wesołowski, J. (2007). Multiple Wiener-Itô integrals as weak limits for p -statistics of increasing orders. *Manuscript under review.*

Resnick, S. I. (1973). Limit laws for record values. *Stochastic Processes Appl.*, 1:67–82.

Rubin, H. and Vitale, R. A. (1980). Asymptotic distribution of symmetric statistics. *Ann. Statist.*, 8(1):165–170.

Ryser, H. J. (1963). *Combinatorial mathematics.* Published by The Mathematical Association of America.

Serfling, R. J. (1980). *Approximation theorems of mathematical statistics.* John Wiley & Sons Inc., New York. Wiley Series in Probability and Mathematical Statistics.

Sinha, K., Kageyama, S., and Singh, M. K. (1993). Construction of rectangular designs. *Statistics*, 25(1):63–70.

Sinha, K. and Mitra, R. K. (1999). Construction of nested balanced block designs, rectangular designs and q-ary codes. *Ann. Comb.*, 3(1):71–80. Combinatorics and biology (Los Alamos, NM, 1998).

Stahly, G. F. (1976). A construction for pbibd(2)'s. *J. Combinatorial Theory Ser. A*, 21(2):250–252.

Stinson, D. R. (1984). A short proof of the nonexistence of a pair of orthogonal Latin squares of order six. *J. Combin. Theory Ser. A*, 36(3):373–376.

Street, A. P. and Street, D. J. (1987). *Combinatorics of experimental design*. The Clarendon Press Oxford University Press, New York.

Székely, G. J. (1982). A limit theorem for elementary symmetric polynomials of independent random variables. *Z. Wahrsch. Verw. Gebiete*, 59(3):355–359.

Taylor, R. L., Daffer, P. Z., and Patterson, R. F. (1985). *Limit theorems for sums of exchangeable random variables*. Rowman & Allanheld Publishers, Totowa, N.J.

Teicher, H. (1988). Distribution and moment convergence of martingales. *Probab. Theory Related Fields*, 79(2):303–316.

Valiant, L. G. (1979). The complexity of computing the permanent. *Theoret. Comput. Sci.*, 8(2):189–201.

van Es, A. J. and Helmers, R. (1988). Elementary symmetric polynomials of increasing order. *Probab. Theory Related Fields*, 80(1):21–35.

Vartak, M. N. and Diwanji, S. M. (1989). Construction of some classes of column-regular BTDs, GDDs and 3-PBIBDs with the rectangular association scheme. *Ars Combin.*, 27:19–39.

Vitale, R. A. (1992). Covariances of symmetric statistics. *J. Multivariate Anal.*, 41(1):14–26.

Wiener, N. (1938). The Homogeneous Chaos. *Amer. J. Math.*, 60(4):897–936.

Wilkinson, D. J. (2006). *Stochastic modelling for systems biology*. Chapman & Hall/CRC Mathematical and Computational Biology Series. Chapman & Hall/CRC, Boca Raton, FL.

Index

1997–1998	Emerging Applications of Dynamical Systems
1998–1999	Mathematics in Biology
1999–2000	Reactive Flows and Transport Phenomena
2000–2001	Mathematics in Multimedia
2001–2002	Mathematics in the Geosciences
2002–2003	Optimization
2003–2004	Probability and Statistics in Complex Systems: Genomics, Networks, and Financial Engineering
2004–2005	Mathematics of Materials and Macromolecules: Multiple Scales, Disorder, and Singularities
2005–2006	Imaging
2006–2007	Applications of Algebraic Geometry
2007–2008	Mathematics of Molecular and Cellular Biology
2008–2009	Mathematics and Chemistry

IMA SUMMER PROGRAMS

1987	Robotics
1988	Signal Processing
1989	Robust Statistics and Diagnostics
1990	Radar and Sonar (June 18–29)
	New Directions in Time Series Analysis (July 2–27)
1991	Semiconductors
1992	Environmental Studies: Mathematical, Computational, and Statistical Analysis
1993	Modeling, Mesh Generation, and Adaptive Numerical Methods for Partial Differential Equations
1994	Molecular Biology
1995	Large Scale Optimizations with Applications to Inverse Problems, Optimal Control and Design, and Molecular and Structural Optimization
1996	Emerging Applications of Number Theory (July 15–26)
	Theory of Random Sets (August 22–24)
1997	Statistics in the Health Sciences
1998	Coding and Cryptography (July 6–18)
	Mathematical Modeling in Industry (July 22–31)
1999	Codes, Systems, and Graphical Models (August 2–13, 1999)
2000	Mathematical Modeling in Industry: A Workshop for Graduate Students (July 19–28)
2001	Geometric Methods in Inverse Problems and PDE Control (July 16–27)
2002	Special Functions in the Digital Age (July 22–August 2)
2003	Probability and Partial Differential Equations in Modern Applied Mathematics (July 21–August 1)
2004	n-Categories: Foundations and Applications (June 7–18)

2005 Wireless Communications (June 22–July 1)
2006 Symmetries and Overdetermined Systems of Partial Differential
        Equations (July 17–August 4)
2007 Classical and Quantum Approaches in Molecular Modeling
        (July 23–August 3)
2008 Geometrical Singularities and Singular Geometries (July 14–25)

IMA "HOT TOPICS" WORKSHOPS

- Challenges and Opportunities in Genomics: Production, Storage, Mining and Use, April 24–27, 1999
- Decision Making Under Uncertainty: Energy and Environmental Models, July 20–24, 1999
- Analysis and Modeling of Optical Devices, September 9–10, 1999
- Decision Making under Uncertainty: Assessment of the Reliability of Mathematical Models, September 16–17, 1999
- Scaling Phenomena in Communication Networks, October 22–24, 1999
- Text Mining, April 17–18, 2000
- Mathematical Challenges in Global Positioning Systems (GPS), August 16–18, 2000
- Modeling and Analysis of Noise in Integrated Circuits and Systems, August 29–30, 2000
- Mathematics of the Internet: E-Auction and Markets, December 3–5, 2000
- Analysis and Modeling of Industrial Jetting Processes, January 10–13, 2001
- Special Workshop: Mathematical Opportunities in Large-Scale Network Dynamics, August 6–7, 2001
- Wireless Networks, August 8–10, 2001
- Numerical Relativity, June 24–29, 2002
- Operational Modeling and Biodefense: Problems, Techniques, and Opportunities, September 28, 2002
- Data-driven Control and Optimization, December 4–6, 2002
- Agent Based Modeling and Simulation, November 3–6, 2003
- Enhancing the Search of Mathematics, April 26–27, 2004
- Compatible Spatial Discretizations for Partial Differential Equations, May 11–15, 2004
- Adaptive Sensing and Multimode Data Inversion, June 27–30, 2004
- Mixed Integer Programming, July 25–29, 2005
- New Directions in Probability Theory, August 5–6, 2005
- Negative Index Materials, October 2–4, 2006
- The Evolution of Mathematical Communication in the Age of Digital Libraries, December 8–9, 2006
- Math is Cool! and Who Wants to Be a Mathematician?, November 3, 2006
- Special Workshop: Blackwell-Tapia Conference, November 3–4, 2006
- Stochastic Models for Intracellular Reaction Networks, May 11–13, 2008

SPRINGER LECTURE NOTES FROM THE IMA:

The Mathematics and Physics of Disordered Media
    Editors: Barry Hughes and Barry Ninham
    (Lecture Notes in Math., Volume 1035, 1983)

Orienting Polymers
    Editor: J.L. Ericksen
    (Lecture Notes in Math., Volume 1063, 1984)

New Prespectives in Thermodynamics
    Editor: James Serrin
    (Springer-Verlag, 1986)

Models of Economic Dynamics
    Editor: Hugo Sonnenschein
    (Lecture Notes in Econ., Volume 264, 1986)

The IMA Volumes in Mathematics and its Applications

The full list of IMA books can be found at the Web site of Institute for
    Mathematics and its Applications:
      http://www.ima.umn.edu/springer/volumes.html

Printed in the United States of America